Stochastic Differential Equations:
Theory and Applications

Stochastic Differential Equations:

Theory and Applications

LUDWIG ARNOLD

A WILEY-INTERSCIENCE PUBLICATION

JOHN WILEY & SONS, New York • London • Sydney • Toronto

Original German language edition published
by R. Oldenbourg Verlag, Munich.
Copyright © 1973 by R. Oldenbourg Verlag.

Copyright © 1974, by John Wiley & Sons, Inc.

Library of Congress Cataloging in Publication Data:

Arnold, Ludwig, 1937–
Stochastic differential equations.

"A Wiley–Interscience publication."
Translation of Stochastische Differentialgleichengen.
Bibliography: p.
1.   Stochastic differential equations.   I.   Title.

QA274.23.A7713   519.2   73-22256
ISBN 0-471-03359-6

Printed in the United States of America

10 9 8 7 6 5 4 3 2 1

Preface

The theory of stochastic differential equations was originally developed by mathematicians as a tool for explicit construction of the trajectories of diffusion processes for given coefficients of drift and diffusion. In the physical and engineering sciences, on the other hand, stochastic differential equations arise in a quite natural manner in the description of systems on which so-called "white noise" acts.

As a result of this variety in the motivations, the existing detailed studies of the subject, as a rule, either are not written from a standpoint of applications or are inaccessible to the person intending to apply them. This holds for the important original work of Itô [42] as well as the books by Gikhman and Skorokhod [5], Dynkin [21], McKean [45], and Skorokhod [47]. The shorter accounts of stochastic dynamic systems in books on stability, filtering, and control (see, for example, Bucy and Joseph [61], Khas'minskiy [65], Jazwinski [66], or Kushner [72]) are rather unsuited for the study and understanding of the subject. Because of the language barrier, the comprehensive book by Gikhman and Skorokhod [36] in Russian is likely to be inaccessible to most persons interested in the subject. (Note added during printing: An English translation has just been published by Springer Verlag.) Also, to a great extent it deals with a considerably more general case.

This book is based on a course that I gave in the summer semester of 1970 at the University of Stuttgart for fifth-semester students of mathematics and engineering. It is written at a moderately advanced level. Apart from probability theory, the only prerequisite is the mathematical preparation usual for students of physical and engineering sciences. The reader can obtain for himself the necessary familiarity with probability theory from any one of a large number of good texts (see [1]-[17]). I have summarized the most important concepts and results of probability theory in Chapter 1, though this is intended only for reference and review purposes. It is only with an intuitive understanding of the basic concepts of probability theory that the reader will be able to distinguish between methodological considerations and technical details.

Throughout, I have gone to great pains (by means of remarks, examples, special cases, and mention of heuristic considerations) to achieve the greatest clarity possible. Proofs that do not provide much information for the development of the subject proper have been omitted. The chapters on stability, filtering, and control include a few examples of possible uses of this new tool.

Professor Peter Sagirow of the University of Stuttgart suggested that I write this book. He also read the manuscript and made many valuable suggestions for improvement. For this I wish to express my gratitude to him. I also thank sincerely Hede Schneider of Stuttgart and Dominique Gaillochet of Montreal for typing the manuscript. Finally, I wish to express my thanks to the Centre de Récherches Mathématiques of the University of Montreal, where I was able to complete the manuscript during my stay there in the academic year 1970–1971.

Ludwig Arnold

Montreal, Quebec
March 1971

Contents

Introduction

Differential equations for random functions (stochastic processes) arise in the investigation of numerous physics and engineering problems. They are usually of one of the following two fundamentally different types.

On the one hand, certain functions, coefficients, parameters, and boundary or initial values in classical differential equation problems can be random. Simple examples are

$$\dot{X}_t = A(t) X_t + B(t), \qquad X_{t_0} = c,$$

with random functions $A(t)$ and $B(t)$ as coefficients and with random initial value c or

$$\dot{X}_t = f(t, X_t, \eta_t), \qquad X_{t_0} = c,$$

with the random function η_t, the random initial value c, and the fixed function f (all the functions are scalar). If these random functions have certain regularity properties, one can consider the above-mentioned problems simply as a family of classical problems for the individual sample functions and treat them with the classical methods of the theory of differential equations.

The situation is quite different if random "functions" of the so-called "white noise" type appear in what is written formally as an ordinary differential equation, for example, the "function" ξ_t in the equation

(a) $$\dot{X}_t = f(t, X_t) + G(t, X_t) \xi_t, \qquad X_{t_0} = c.$$

This "white noise" is conceived as a stationary Gaussian stochastic process with mean value zero and a constant spectral density on the entire real axis. Such a process does not exist in the conventional sense, since it would have to have the Dirac delta function as covariance function, and hence an infinite variance (and independent values at all points). Nonetheless, the "white noise" ξ_t is a very useful mathematical idealization for describing random influences that fluctuate rapidly and hence are virtually uncorrelated for different instants of time.

Such equations were first treated in 1908 by Langevin [44] in the study of the Brownian motion of a particle in a fluid. If X_t is a component of the velocity, at an instant t, of a free particle that performs a Brownian motion, Langevin's equation is

(b) $$\dot{X}_t = -\alpha X_t + \sigma \xi_t, \qquad \alpha > 0, \ \sigma \text{ constants.}$$

Here, $-\alpha X_t$ is the systematic part of the influence of the surrounding medium due to dynamic friction. The constant α is found from Stokes's law to be

$\alpha = 6 \pi a \eta/m$, where a is the radius of the (spherical) particle, m is its mass, and η is the viscosity of the surrounding fluid. On the other hand, the term $\sigma \xi_t$ represents the force exerted on the particle by the molecular collisions. Since under normal conditions the particle uniformly undergoes about 10^{21} molecular collisions per second from all directions, $\sigma \xi_t$ is indeed a rapidly varying fluctuational term, which can be idealized as "white noise." If we normalize ξ_t so that its covariance is the delta function, then $\sigma^2 = 2 \alpha k T/m$ (where k is Boltzmann's constant and T is the absolute temperature of the surrounding fluid). The same equation (b) arises formally for the current in an electric circuit. This time, ξ_t represents the thermal noise. Of course, (b) is a special case of equation (a), the right–hand member of which is decomposed as the sum of a systematic part f and a fluctuational part $G \xi_t$.

In model (b) of Brownian motion, one can calculate explicitly the probability distributions of X_t even though ξ_t is not a random function in the usual sense. As a matter of fact, every process X_t with these distributions (Ornstein–Uhlenbeck process) has sample functions that, with probability 1, are nondifferentiable, so that (b) and, more generally, (a) cannot be regarded as ordinary differential equations.

For a mathematically rigorous treatment of equations of type (a), a new theory is necessary. This is the subject of the present book. It turns out that, whereas "white noise" is only a generalized stochastic process, the indefinite integral

(c) $$W_t = \int_0^t \xi_s \, ds$$

can nonetheless be identified with the Wiener process. This is a Gaussian stochastic process with continuous (but nowhere differentiable) sample functions, with mean $E W_t = 0$ and with covariance $E W_t W_s = \min (t, s)$.

If we write (c) symbolically as

$$dW_t = \xi_t \, dt,$$

(a) can be put in the differential form

(d) $$dX_t = f (t, X_t) \, dt + G (t, X_t) \, dW_t, \qquad X_{t_0} = c.$$

This is a stochastic differential equation (Itô's) for the process X_t. It should be understood as an abbreviation for the integral equation

(e) $$X_t = c + \int_{t_0} f (s, X_s) \, ds + \int_{t_0}^t G (s, X_s) \, dW_s.$$

Since the sample functions of W_t are with probability 1 continuous though not of bounded variation in any interval, the second integral in (e) cannot, even for smooth G, be regarded in general as an ordinary Riemann–Stieltjes integral with respect to the sample functions of W_t because the value depends on the intermediate points in the approximating sums. In 1951, Itô [42] defined integrals of the form

$$Y_t = \int_{t_0}^{t} G(s)\, dW_s$$

for a broad class of so-called nonanticipating functionals G of the Wiener process W_t and in doing so put the theory of stochastic differential equations on a solid foundation. This theory has its peculiarities. For example, the solution of the equation

$$dX_t = X_t\, dW_t, \qquad X_0 = 1,$$

is not e^{W_t}, but

$$X_t = e^{W_t - t/2},$$

which one does not derive by purely formal calculation according to the classical rules. It turns out that the solution of a stochastic differential equation of the form (d) is a Markov process with continuous sample functions—in fact, a diffusion process. Conversely, every (smooth) diffusion process is the solution of a stochastic differential equation of the form (d) where f and G^2 are respectively the coefficients of drift and diffusion.

For diffusion processes, there exist effective methods for calculating transitional and finite-dimensional distributions and distributions of many functionals. These methods belong to the so-called analytical or indirect of probability methods which deal not with the timewise development of the state X_t, but, for example, with the timewise development of transition probabilities $P(X_t \in B | X_s = x)$.

In contrast, the calculus of stochastic differential equations belongs to the probabilistic or direct methods since with them we are concerned with the random variable X_t and its variation. An equation of the form (d) or (e) represents a construction rule (albeit in general a complicated one) with which one can construct the trajectories of X_t from the trajectories of a Wiener process W_t and an initial value c.

The law of motion for state of a stochastic dynamic system without after-effects ("without memory") can be described by an equation more general than (d), namely, by an equation of the form

(f) $$dX_t = g(t, X_t, dt).$$

In the case of fluctuational influences that are additively superimposed on a systematic part, we have

$$g(t, x, h) = f(t, x)\, h + G(t, x)\, (Y_{t+h} - Y_t).$$

Here, Y_t is a process with independent increments and equation (f) takes the form

$$dX_t = f(t, X_t)\, dt + G(t, X_t)\, dY_t.$$

Such equations have been studied by Itô [42]. We shall confine ourselves here to the most important special case $Y_t = W_t$.

Notation and Abbreviations

Vectors of dimension d are basically treated as $d \times 1$ matrices (column vectors). Equations or inequalities involving random variables hold in general only with probability 1. As a rule, the argument ω of random variables is omitted.

A' Transpose of the matrix A

X_t A d-dimensional stochastic process with index set $[t_0, T] \subset [0, \infty) = R^+$

X'_t The transpose (vector) of X_t

$\dot{X}_t$ The derivative of X_t with respect to t

R^d d-dimensional Euclidean space with distance function $|x - y|$

I Unit matrix

I_A The indicator function of the set A

$|x|$ The norm of $x \in R^d$: $|x|^2 = \sum\limits_{i=1}^{d} x_i^2 = x' x = \mathrm{tr}\,(x\,x')$

$x'\,y$ Scalar product of $x, y \in R^d$, $x' y = \sum\limits_{i=1}^{d} x_i y_i = \mathrm{tr}\,(x\,y')$

$x\,y'$ Matrix $(x_i\,y_j)$

$|A|$ Norm of the $d \times m$ matrix A : $|A|^2 = \sum\limits_{i=1}^{d} \sum\limits_{j=1}^{m} a_{ij}^2 = \mathrm{tr}\,A\,A'$ (note that

$|A\,x| \leqq |A|\,|x|, \;\; |A\,B| \leqq |A|\,|B|$

$\mathrm{tr}\,A = \sum\limits_{i=1}^{d} a_{ii}$ Trace of the matrix A

A positive-definite (resp. nonnegative-definite): $x'\,A\,x > 0$ (resp. $\geqq 0$) for all $x \neq 0$

$\delta\,(t)$ Dirac's delta function

δ_x Probability measure concentrated at the point x

sup (resp. inf) Least upper (greatest lower) bound of a scalar set or sequence

lim sup (resp. lim inf) Greatest (resp. least) point of accumulation of a scalar sequence

$o\,(g\,(t))$, $O\,(g\,(t))$ Quantity whose ratio to $g\,(t)$ (usually as $t \to 0$) approaches 0 (resp. remains bounded).

$(\Omega, \mathfrak{A}, P)$ Probability space

$\mathfrak{A}\,(\mathfrak{C})$ The sigma–algebra generated by the family $\mathfrak{C}$ of sets

$\mathfrak{B}^d$, $\mathfrak{B}^d\,(M)$ The sigma–algebra of Borel sets in R^d (resp. in $M \subset R^d$)

$\mathfrak{A}\,([t_1, t_2])$ The sigma–algebra generated by the random variables X_t for $t_1 \leqq t \leqq t_2$

ξ_t A (vector-valued) white noise

W_t A (vector-valued) Wiener process

$L^p = L^p\,(\Omega, \mathfrak{A}, P)$ All random variables such that $E\,|X|^p < \infty$

$X_{.}\,(\omega)$, $f\,(\cdot, x)$ X (resp. f) as a function of the variables replaced by the dot for fixed ω (resp. x)

$L_2\,[t_0, T]$ All measurable functions such that $\displaystyle\int_{t_0}^{T} |f\,(s)|^2\,\mathrm{d}s < \infty$

$P\,(s, x, t, B)$ Probability of transition of the point x at the instant s into the set B at the instant t

$p\,(s, x, t, y)$ Density of $P\,(s, x, t, \cdot)$

$n\,(t, x, y) = (2\,\pi\,t)^{-d/2}\,e^{-|y-x|^2/2t}$

$\mathfrak{N}\,(m, C)$ d -dimensional normal distribution with expectation vector m and covariance matrix C

st–lim (resp. qm–lim, resp. ac–lim) Stochastic limit (resp. mean square or quadratic mean limit, resp. limit with probability 1) of a sequence of random variables

Stochastic Differential Equations: Theory and Applications

Chapter 1

Fundamentals of Probability Theory

1.1 Events and Random Variables

Probability theory deals with mathematical models of trials whose outcomes depend on chance. We group together the possible outcomes—the elementary events—in a set Ω with typical element $\omega \in \Omega$. If the trial is the tossing of a coin, then $\Omega = \{$heads, tails$\}$; for the throwing of a pair of (distinguishable) dice, $\Omega = \{(i, j): 1 \leq i, j \leq 6\}$; for the life length of a light bulb, $\Omega = [0, \infty)$; in the observation of water level from time t_1 to time t_2, Ω is the set of all real functions (or perhaps all continuous functions) defined on the interval $[t_1, t_2]$. An observable **event** A is a subset of Ω, which we indicate by writing $A \subset \Omega$. (In the dice example, A might be $\{(i, j): i + j = $ an even number$\}$, and in the light bulb example, A might be $\{\omega: \omega \geq c\}$.)

On the other hand, not every subset of Ω is in general an observable or interesting event. Let $\mathfrak{A}$ denote the set of observable events for a single trial. Of course, $\mathfrak{A}$ must include the certain event Ω, the impossible event $\emptyset$ (the empty set) and, for every event A, its complement $\overline{A}$. Furthermore, given two events A and B in $\mathfrak{A}$, the union $A \cup B$ and the intersection $A \cap B$ also belong to $\mathfrak{A}$; thus, $\mathfrak{A}$ is an **algebra** of events. In many practical problems, one must be able to make countable unions and intersections in $\mathfrak{A}$. To do this, it is sufficient to assume that

$$\bigcup_{n=1}^{\infty} A_n \in \mathfrak{A}$$

when $A_n \in \mathfrak{A}$ for $n \geq 1$. An algebra $\mathfrak{A}$ of events with this property is called a **sigma algebra**. Henceforth, we shall deal with sigma algebras exclusively. In the terminology of measure theory, which is parallel to the terminology of probability theory, the elements of $\mathfrak{A}$ are called **measurable sets** and the pair $(\Omega, \mathfrak{A})$ is called a **measurable space**. Two events A and B are said to be **incompatible** if they are disjoint, that is, if $A \cap B = \emptyset$. If A is a subset of B, indicated by writing $A \subset B$ (where $A = B$ is allowed), we say that A **implies** B.

Let $\mathfrak{C}$ denote a family of subsets of Ω. Then, there exists in Ω a smallest sigma-algebra $\mathfrak{A}(\mathfrak{C})$ that contains all sets belonging to $\mathfrak{C}$. This $\mathfrak{A}(\mathfrak{C})$ is called the **sigma algebra generated by** $\mathfrak{C}$.

Let $(\Omega, \mathfrak{A})$ and $(\Omega', \mathfrak{A}')$ denote measurable spaces. A mapping $X: \Omega \to \Omega'$ that assigns to every $\omega \in \Omega$ a member $\omega' = X(\omega)$ of Ω' is said to be $(\mathfrak{A}\text{-}\mathfrak{A}')$-**measurable** (and is called an Ω'-valued **random variable** on $(\Omega, \mathfrak{A})$) if the preimages of measurable sets in Ω' are measurable sets in Ω, that is, if, for $A' \in \mathfrak{A}'$,

$$\{\omega: X(\omega) \in A'\} = [X(\omega) \in A'] = X^{-1}(A') \in \mathfrak{A}.$$

The set $\mathfrak{A}(X)$ of preimages of measurable sets is itself a sigma algebra in Ω and is the smallest sigma algebra with respect to which X is measurable. It is called the **sigma algebra generated by** X **in** Ω.

Suppose that Ω' is the d-dimensional Euclidean space R^d with distance function

$$|x - y| = \left(\sum_{k=1}^{d} |x_k - y_k|^2 \right)^{1/2}, \qquad x = \begin{pmatrix} x_1 \\ x_2 \\ \vdots \\ x_d \end{pmatrix}, \quad y = \begin{pmatrix} y_1 \\ y_2 \\ \vdots \\ y_d \end{pmatrix}.$$

In this special case, we shall always choose as the sigma algebra $\mathfrak{A}'$ of events the **sigma algebra $\mathfrak{B}^d$ of Borel sets** in R^d generated by the d-dimensional intervals $a_k \leq x_k \leq b_k$ for $k = 1, 2, \ldots, d$. Borel sets include in particular all closed and all open sets and all kinds of d-dimensional intervals (half-open, unbounded, etc.). Although there are "many" non-Borel sets, it is not easy to exhibit specific examples of them. If Ω' is a subset of R^d (examples for $d = 1$ are $[0, \infty), [0, 1], \{0, 1, 2, \ldots\}$, etc.), then we always choose $\mathfrak{A}' = \mathfrak{B}^d(\Omega') = \{A' = B \cap \Omega': B \in \mathfrak{B}^d\}$.

An R^d-valued function $X: \Omega \to R^d$, where $(\Omega, \mathfrak{A})$ is a measurable space, is measurable (resp. Borel-measurable, resp. $\mathfrak{A}$-measurable) if and only if the d components X_k, for $1 \leq k \leq d$, are real-valued (or scalar) measurable (resp. Borel-measurable, resp. $\mathfrak{A}$-measurable) functions. A real-valued function $X: \Omega \to R^1$ is measurable if and only if the preimages of all intervals of the form $(-\infty, a]$ are measurable, that is, if and only if

$$\{\omega: X(\omega) \leq a\} = [X \leq a] \in \mathfrak{A} \quad \text{for all } a \in R^1.$$

A $(d \times m)$-matrix-valued function

$$X(\omega) = \begin{pmatrix} X_{11}(\omega) & \ldots & X_{1m}(\omega) \\ \vdots & & \vdots \\ X_{d1}(\omega) & \ldots & X_{dm}(\omega) \end{pmatrix}$$

is measurable (as a function with range in R^{dm}) if and only if the elements X_{ij} are measurable.

The **indicator function** I_A of a set $A \subset \Omega$ is defined by

$$I_A(\omega) = \begin{cases} 1 & \text{for} \quad \omega \in A, \\ 0 & \text{for} \quad \omega \notin A. \end{cases}$$

The indicator function I_A is measurable if and only if A is measurable, that is, $A \in \mathfrak{A}$.

Calculation with measurable functions is greatly facilitated by the fact that the set of real-valued measurable functions is closed under the operations of classical analysis, that is, under at most countably infinite applications of addition, subtraction, multiplication, division, and evaluation of the supremum, infimum, and limits superior and inferior provided the operations lead to another real-valued function. In particular, the limit of a sequence of measurable functions is a measurable function.

The intuitive background of the concept of a random variable or measurable function is as follows: Suppose that we are given a trial described by a measurable space $(\Omega, \mathfrak{A})$, where Ω is the set of possible observable elementary events and $\mathfrak{A}$ is the sigma algebra of events that are observable or interesting in the framework of the trial.

Now it could happen that we would not be able to observe ω directly but, with the help of a measuring instrument, could measure a value $X(\omega)$ in a set Ω', where the sigma algebra of events $\mathfrak{A}'$ is defined on Ω'. Thus, the value of $X(\omega)$ would depend on ω, that is, on chance. The assumption of measurability of the function X means that for every observable meaningful event in the space Ω' there is a meaningful event in the original space. Now, information is, in general, lost by observing $X(\omega)$ instead of ω, a fact expressed by the condition that $\mathfrak{A}(X)$ is only a sub-sigma-algebra of $\mathfrak{A}$. In throwing of a pair of dice, we can choose in $\Omega = \{(i, j): 1 \leq i, j \leq 6\}$ the set of all subsets of Ω to serve as $\mathfrak{A}$. Every function defined on Ω is then measurable. On the other hand, if $\mathfrak{A}$ is the sigma algebra generated by the sets $A_k = \{(i, j): i + j = k\}$, where $2 \leq k \leq 12$ (that is, if only the sum of the spots can be observed), then, for example, $X((i, j)) = i$ is not measurable.

1.2 Probability and Distribution Functions

Let $(\Omega, \mathfrak{A})$ denote a measurable space. A set function μ defined on $\mathfrak{A}$ is called a **measure** if

a) $0 \leq \mu(A) \leq \infty$ for all $A \in \mathfrak{A}$,

b) $\mu(\emptyset) = 0$,

c)

$$\mu \left(\bigcup_{n=1}^{\infty} A_n \right) = \sum_{n=1}^{\infty} \mu \left(A_n \right) \quad \text{if} \quad A_n \in \mathfrak{A} \quad \text{for all}$$

$n \geqq 1$ and $A_n \cap A_m = \emptyset \; (n \neq m)$ (sigma-additivity).

The triple $(\Omega, \mathfrak{A}, \mu)$ is called a **measure space**, and $\mu (A)$ is called the measure of the set A. If Ω is the union of an at most countably infinite family of sets in $\mathfrak{A}$, each with finite measure, then μ is said to be **sigma-finite**. If $\mu (\Omega) < \infty$, then μ is said to be **finite**. A normed finite measure P, that is, a measure P with the property

$$P (\Omega) = 1,$$

is called a probability measure or simply a **probability**, and the triple $(\Omega, \mathfrak{A}, P)$ is called a **probability space**. The set function P thus assigns to every event A a number $P (A)$, known as the probability of A, such that $0 \leqq P (A) \leqq 1$.

Furthermore, we have

$$P (A) = 1 - P (\overline{A}), \quad A \in \mathfrak{A},$$

and

$$P (A) \leqq P (B) \quad \text{for} \quad A \subset B, \quad A, B \in \mathfrak{A}.$$

Probability theory is concerned, roughly speaking, only with the calculation of new probabilities from given ones. The *a priori* probabilities are obtained either from theoretical considerations or from observation of frequencies in long series of trials. We shall use the *frequency interpretation* of probability theory as the basis of our discussion. This interpretation means that $P (A)$ is the theoretical value of the relative frequency of A in a large number n of performances of a trial characterized by the measurable space $(\Omega, \mathfrak{A})$; that is,

$$P (A) \approx \frac{n_A}{n},$$

where n is the number of performances of the trial and n_A is the number of occurrences of the event A in the course of the n performances. Estimations of $P (A)$ are provided by statistics.

When a pair of dice are to be thrown, a probability is assigned to all subsets of $\Omega = \{(i, j): 1 \leqq i, j \leqq 6\}$ by assigning (and this holds for all countable sets Ω) to the singleton sets $\{(i, j)\}$ the specific probabilities:

$$P (\{(i, j)\}) = p_{ij} \geqq 0, \quad \sum_{i=1}^{6} \sum_{j=1}^{6} p_{ij} = 1.$$

Then, for every set A,

$$P (A) = \sum_{(i, j) \in A} \sum p_{ij}.$$

If the dice are independent (see section 1.5), the p_{ij} have the form

$$p_{ij} = p_i\, q_j, \quad \sum_{i=1}^{6} p_i = \sum_{j=1}^{6} q_j = 1.$$

Finally, if the dice are "fair", then $p_i = q_i = 1/6$, so that $p_{ij} = 1/36$.

Of all nonnormed measures, we are interested here only in **Lebesgue measure** λ, defined on the set of Borel sets $\mathfrak{B}^d$ in R^d. This measure assigns to every d-dimensional interval its "length":

$$\lambda\left(\{x\colon a_k \leqq x_k \leqq b_k\}\right) = \prod_{k=1}^{d} (b_k - a_k).$$

Thus, in the case of simple sets, it corresponds to their elementary geometric content and it can be carried over in an unambiguous way from intervals to all Borel sets. Since $\lambda(R^d) = \infty$, Lebesgue measure is not finite. However, since

$$\lambda\left(\{x\colon -n \leqq x_k \leqq n\}\right) = (2\,n)^d < \infty, \quad n = 1, 2, \dots,$$

it is a sigma-finite measure. Every countable set (for example, the set of points with rational coordinates) has Lebesgue measure 0.

Let $(\Omega, \mathfrak{A}, \mu)$ denote a measure space and let $E(\omega)$ denote a proposition regarding the elements ω of Ω. Then, we shall write $E[\mu]$ to mean that E is true for all ω in Ω except possibly for those ω in some set N (belonging to $\mathfrak{A}$) such that $\mu(N) = 0$, so that E, is true for all $\omega \in \bar{N}$ is a probability P, we shall say "certain $[P]$" or, since $P(N) = 1$, "with probability 1".

Now, suppose that $(\Omega, \mathfrak{A}, P)$ is a probability space, that $(\Omega', \mathfrak{A}')$ is a measurable space, and that X is a random variable with values in Ω'. The function X maps the probability P onto the measurable space of the images by

$$P_X(A') = P(X^{-1}(A')) = P\{\omega\colon X(\omega) \in A'\} = P[X \in A'] \quad \text{for all}$$
$$A' \in \mathfrak{A}'.$$

The function P_X is called the **distribution** of X. It contains the information needed for probability-theoretic examination of X. For an R^d-valued random variable, the distribution P_X is uniquely defined on $\mathfrak{B}^d$ by its **distribution function**

$$F(x) = F(x_1, \dots, x_d) = P\{\omega\colon X_1(\omega) \leqq x_1, \dots, X_d(\omega) \leqq x_d\}$$
$$= P[X \leqq x],$$

which shows how likely it is that X will assume values to the "left" of the point $x \in R^d$. The function $F(x)$ is a convenient tool for describing the distribution of X inasmuch as it is not a set function but an ordinary point function defined on R^d. It is also called the **joint distribution function** of the d scalar random variables $X_1, \dots, X_d$, which are the components of X. For $d = 1$, $F(x)$ is an increasing function that is everywhere continuous from the right; also,

$$F(-\infty) = \lim_{x \to -\infty} F(x) = 0, \quad F(\infty) = \lim_{x \to \infty} F(x) = 1.$$

These properties carry over in an obvious way to the case $d \geq 2$. Every function with these properties is called a distribution function. In applications, random variables are defined in terms of their distribution functions. In fact, for given F, it is always possible to construct a probability space $(\Omega, \mathfrak{A}, P)$ and a random variable X in such a way that F is the distribution function of X. For example, one might choose $(\Omega, \mathfrak{A}, P) = (R^d, \mathfrak{B}^d, P_F)$ and $X(\omega) \equiv \omega$, where P_F is the probability uniquely defined on $\mathfrak{B}^d$ by F.

If F is the distribution function of an R^d-valued random variable X, it is possible to obtain the distribution function (**marginal distribution**) of a group $(X_{n_1}, \dots, X_{n_k})$ of k components (where $k \leq d$) of X by replacing in $F(x_1, \dots, x_d)$ the unencountered arguments with ∞. For example, we obtain the one-dimensional marginal distributions as follows:

$$F_k(x_k) = P[X_k \leq x_k] = F(\infty, \dots, \infty, x_k, \infty, \dots, \infty).$$

If the probability P_F generated on $\mathfrak{B}^d$ by F is Lebesgue-continuous (that is, if $P_F(N) = 0$ for every set of Lebesgue measure zero belonging to $\mathfrak{B}^d$), then P_F has a density (see section 1.3); that is, there exists an integrable function $f(x) \geq 0$ such that

$$F(x_1, \dots, x_d) = \int_{-\infty}^{x_1} \dots \int_{-\infty}^{x_d} f(y_1, \dots, y_d)\, dy_1 \dots dy_d.$$

The distribution function F is then **absolutely continuous** (and hence continuous) in the classical sense; that is, for every $\varepsilon > 0$, there exists a δ such that, for finitely many disjoint intervals $I_1, \dots, I_n$, the total increase in F on these intervals is less than ε whenever their total length is less than δ. Since F is an absolutely continuous function, it is differentiable $[\lambda]$, and

$$\frac{\partial^d F}{\partial x_1 \, \partial x_2 \dots \partial x_d} = f(x_1, x_2, \dots, x_d). \ [\lambda]$$

In particular, equality holds wherever the density f is continuous.

An R^d-valued random variable has a **normal distribution** (a **Gaussian distribution**) $\mathfrak{N}(m, C)$, where $m \in R^d$ and C is a positive-definite $d \times d$ matrix, whenever its distribution function has density

$$f(x) = ((2\pi)^d \det(C))^{-1/2} \exp\left(-\frac{1}{2}(x-m)' \, C^{-1}(x-m)\right).$$

The probability-theoretic significance of the parameters m and C will be made clear in the next section.

1.3 Integration Theory, Expectation

To a real-valued random variable X defined on $(\Omega, \mathfrak{A}, P)$ we now wish to assign a specific number, namely its expectation. If X assumes only finitely many distinct values $c_1, \ldots, c_n$, the expectation is defined by

$$E X = \sum_{i=1}^{n} c_i P [X = c_i].$$

This is a special case of an integral which by an approximation process can be extended to arbitrary real-valued random variables.

Let us summarize integration theory for real-valued random variables over an arbitrary space $(\Omega, \mathfrak{A}, P)$. An integral is defined in three steps:

Step 1. Suppose that

$$X = \sum_{i=1}^{n} c_i I_{A_i}, \quad \bigcup_{i=1}^{n} A_i = \Omega, \quad A_i \cap A_j = \emptyset \, (i \neq j), \quad A_i \in \mathfrak{A},$$

is a so-called **step function**. We define

$$\int_{\Omega} X(\omega) \, dP(\omega) = \int_{\Omega} X \, dP = \sum_{i=1}^{n} c_i P(A_i).$$

Step 2. Now, let $X \geq 0$ be any measurable function. There exists an increasing sequence $\{X_n\}$ of nonnegative measurable step functions such that

$$\lim_{n \to \infty} X_n(\omega) = X(\omega) \quad \text{for all } \omega \in \Omega$$

and

$$\lim_{n \to \infty} \int_{\Omega} X_n \, dP = c \leq \infty.$$

Here, c is independent of the special sequence $\{X_n\}$. We define

$$\int_{\Omega} X \, dP = c.$$

Step 3. Now, let X denote any measurable function. We decompose it into positive and negative parts:

$$X = X^+ - X^-, \quad X^+ = X I_{[X \geq 0]}, \quad X^- = -X I_{[X < 0]}.$$

Then we define

$$\int_{\Omega} X \, dP = \int_{\Omega} X^+ \, dP - \int_{\Omega} X^- \, dP,$$

assuming that this does not yield $\infty - \infty$.

A random variable X is said to be $(P-)$ **integrable** if the integral $\int X \, dP$ is finite. In probability theory, the integral is also called the **expectation** and one writes

$$E(X) = E X = \int_{\Omega} X \, dP.$$

For example, for every $A \in \mathfrak{A}$, we have $E I_A = P(A)$.

For R^d- and matrix-valued random variables, we define

$$\int_{\Omega} X \, dP = E X = \begin{pmatrix} E X_1 \\ \vdots \\ E X_d \end{pmatrix}, \quad X = \begin{pmatrix} X_1 \\ \vdots \\ X_d \end{pmatrix},$$

and

$$\int_{\Omega} A \, dP = E A = (E A_{ij}), \quad A = (A_{ij})_{d \times m}$$

respectively. These expectations exist (in the sense of existence of expectation of each component) if and only if $E |X| < \infty$ in the first case and $E |A| < \infty$ in the second. In the remainder of this section, we shall discuss only R^d-valued random variables unless the contrary is stated.

For actual evaluation of expectations (for example, from distribution functions), the following **transformation theorem** is of importance: let $g: R^d \to R^m$ denote a measurable function. Then, for $Y = g(X)$,

$$E Y = \int_{R^d} g(x) \, dF(x),$$

In particular, for $d = m$ and $g(x) \equiv x$,

$$E X = \int_{R^d} x \, dF(x),$$

provided the integral in question exists. Here, F is the distribution function of X and we write $\int g(x) \, dF(x)$ for the integral of g over the probability space $(R^d, \mathfrak{B}^d, P_X)$—the so-called **Lebesgue-Stieltjes integral**. If it exists, it coincides for continuous g with the usual Riemann-Stieltjes integral.

In the case of an integrable discrete random variable that assumes the values $c_i \in R^d$ with probabilities p_i, the last formula yields

$$E X = \sum_{i=1}^{\infty} c_i p_i,$$

and, for an integrable absolutely continuous random variable with density f (see the Radon-Nikodym theorem below),

$$E X = \int_{R^d} x f(x) \, dx = \begin{pmatrix} \int_{R^d} x_1 f(x_1, \ldots, x_d) \, dx_1 \ldots dx_d \\ \vdots \\ \int_{R^d} x_d f(x_1, \ldots, x_d) \, dx_1 \ldots dx_d \end{pmatrix}$$

$$= \begin{pmatrix} \int_{R^1} x_1 f_1(x_1) \, dx_1 \\ \vdots \\ \int_{R^1} x_d f_d(x_d) \, dx_d \end{pmatrix}, \quad f_i \text{ is the density of } X_i.$$

Here, these integrals are in general Lebesgue integrals, which we shall treat at the end of this section.

Suppose that, for $p \geqq 1$,

$$\mathfrak{L}^p = \mathfrak{L}^p(\Omega, \mathfrak{A}, P) = \{X : X \text{ is an } R^d\text{-valued random variable,}$$
$$E|X|^p < \infty\}.$$

We have $\mathfrak{L}^p \subset \mathfrak{L}^q$ (where $p \geqq q$) and $\mathfrak{L}^p$ is a linear space. If in $\mathfrak{L}^p$ we shift to the set L^p of equivalence classes of random variables that with probability 1 coincide and if we set

$$\|X\|_p = (E|X|^p)^{1/p},$$

then, L^p is a Banach space with respect to this norm. In fact, L^2 is a Hilbert space with scalar product

$$(X, Y) = E X' Y = \sum_{i=1}^{d} E X_i Y_i.$$

In L^1, we have $|E X| \leqq E |X|$ and

$$E(\alpha X + \beta Y) = \alpha E X + \beta E Y \quad \text{(Linearity)}$$

For $d = 1$, we have

$$E X \leqq E Y \quad \text{for} \quad X \leqq Y \quad \text{(Monotonicity)}.$$

In addition, we have **Hölder's inequality** (for $p = 2$, **Schwarz's inequality**)

$$|(X, Y)| \leqq \|X\|_p \|Y\|_q \quad (p > 1, \, 1/p + 1/q = 1, \, X \in L^p, \, Y \in L^q),$$

Minkowski's inequality (the **triangle inequality** in L^p)

$$\|X + Y\|_p \leqq \|X\|_p + \|Y\|_p \quad (p \geqq 1, \, X, Y \in L^p),$$

and **Chebyshev's inequality**

$$P\left[|X| \geq c\right] \leq c^{-p} E |X|^p \qquad (p > 0, c > 0).$$

For $d = 1$, the number

$$V(X) = E(X - EX)^2 = \sigma^2(X), \quad X \in L^2,$$

is called the **variance** of X, the number $\sigma = (V(X))^{1/2}$ is called the **standard deviation** of X, the number $E X^k$ (where k is a natural number and X belongs to L^k) is called the kth **moment** of X, the number $E(X - EX)^k$ is called the kth **central moment** of X, and the number

$$\text{Cov}(X, Y) = E(X - EX)(Y - EY)$$

(where X and Y belong to L^2) is called the **covariance** of X and Y. If $\text{cov}(X, Y) = 0$, we say that X and Y are **uncorrelated**. For $d > 1$, the symmetric nonnegative-definite $d \times d$ matrix

$$\text{Cov}(X, Y) = E(X - EX)(Y - EY)'$$
$$= (\text{Cov}(X_i, Y_j))$$

is called the **covariance matrix** of the R^d-valued random variables X and Y. For $\text{Cov}(X, X)$, we write simply $\text{Cov}(X)$. The **characteristic function** of a random variable X or of its distribution function F is

$$\varphi(t) = \varphi_X(t) = E\, e^{it'X} = \int_{R^d} e^{it'x}\, dF(x), \quad t \in R^d.$$

The distribution function F is uniquely defined by φ.

A random variable X that has the normal distribution $\mathfrak{N}(m, C)$ defined in section 1.2 has expectation vector $EX = m$, covariance matrix $\text{Cov}(X) = C$ and characteristic function

$$\varphi(t) = \exp\left(i\, t'\, m - \frac{1}{2}\, t'\, C\, t\right), \quad t \in R^d.$$

This last relation can also serve as a definition of normal distribution, provided C is nonnegative-definite (and the distribution can therefore be concentrated on a certain linear subspace of R^d whose dimension is equal to the rank of C). For $d = 1$, $\mathfrak{N}(m, \sigma^2)$ has central moments

$$E(X - m)^n = \begin{cases} 0, & n \geq 1, \quad \text{odd}, \\ 1 \cdot 3 \cdot 5 \cdot \ldots \cdot (n-1)\, \sigma^n, & n \geq 2, \quad \text{even}. \end{cases}$$

If X has, for $d \geq 1$, the distribution $\mathfrak{N}(m, C)$, A is a fixed $p \times d$ matrix, and a is a member of R^p, then $Y = A X + a$ has the distribution $\mathfrak{N} (A m + a, A C A')$. In particular, every group of components of a normally distributed vector is normally distributed.

Some important convergence theorems dealing with the integration theory summarized here are:

a) **Theorem on monotonic convergence:** If $\{X_n\}$ is an increasing sequence of nonnegative random variables, then

$$\lim_{n \to \infty} E (X_n) = E (\lim_{n \to \infty} X_n).$$

b) **Theorem on dominated convergence:** Let $\{X_n\}$ denote a sequence of random variables such that $X_n \in L^p$, where $p \geqq 1$,

$$\text{st-}\lim_{n \to \infty} X_n = X$$

(see section 1.4), and $|X_n| \leqq Y \in L^p$. Then, $X \in L^p$, $\|X_n - X\|_p \to 0$, and

$$\lim_{n \to \infty} E X_n = E X.$$

This theorem holds, in particular, when almost certain convergence of X_n to X rather than stochastic convergence obtains.

An integration theory can be developed for an arbitrary measure space $(\Omega, \mathfrak{A}, \mu)$, in exactly the same way as in the case $\mu (\Omega) = 1$. However, the relation $\mathfrak{L}^p \subset \mathfrak{L}^q$ for $p \geqq q$ is false in the case $\mu (\Omega) = \infty$. The only infinite measure space that we consider here is the space $(R^d, \mathfrak{B}^d, \lambda)$. The resulting integral is called the **Lebesgue integral**. We write

$$\int\limits_{R^d} f \, d\lambda = \int\limits_{R^d} f (x) \, dx$$

and we define, for every Borel subset B of R^d,

$$\int\limits_{B} f (x) \, dx = \int\limits_{R^d} f I_B \, d\lambda.$$

In the case $d = 1$ and B is $[a, b], (-\infty, b), R^1$, etc., the notations

$$\int\limits_{a}^{b} f (x) \, dx, \quad \int\limits_{-\infty}^{b} f (x) \, dx, \quad \int\limits_{-\infty}^{\infty} f (x) \, dx$$

are the usual ones. The Lebesgue integral is more general than the familiar Riemann integral defined in terms of upper and lower sums, inasmuch as it is defined for more functions. Every bounded Riemann-integrable (for example, every continuous) function defined on a bounded interval is also Lebesgue-integrable and the two integrals coincide. The same holds for nonnegative integrands and for improper Riemann integrals. All the integrals over R^d or subsets of it that we have been considering are basically Lebesgue integrals though, for the most part, they might, by virtue of the smoothness of the integrands, be regarded as Riemann integrals.

Let ν and μ denote any two measures on $(\Omega, \mathfrak{A})$. The measure ν is said to be μ-continuous if $\nu(N) = 0$ whenever $\mu(N) = 0$. Also, ν is said to have μ-density $f \geqq 0$ whenever

$$\nu(A) = \int_A f \, d\mu = \int_\Omega f \, I_A \, d\mu \quad \text{for all } A \in \mathfrak{A}.$$

We then write

$$f = \frac{d\nu}{d\mu}.$$

Radon-Nikodym theorem: Let ν and μ denote two measures defined on $(\Omega, \mathfrak{A})$, and suppose that μ is sigma-finite. Then, ν is μ-continuous if and only if ν has a μ-density. This density is uniquely defined $[\mu]$. We then have

$$\int_\Omega X \, d\nu = \int_\Omega X \, \frac{d\nu}{d\mu} \, d\mu$$

as long as one side of this equation is meaningful.

If $(\Omega, \mathfrak{A}) = (R^d, \mathfrak{B}^d)$ and $\mu = \lambda$, we also speak of Lebesgue-continuity. If $\nu = P$ is a Lebesgue-continuous probability on $(R^d, \mathfrak{B}^d)$ with distribution function F, then there exists a density $f \geqq 0$ uniquely defined $[\lambda]$ such that

$$P(B) = \int_B f(x) \, dx \quad \text{for all } B \in \mathfrak{B}^d,$$

and we have $[\lambda]$ (in particular, wherever f is continuous)

$$\frac{\partial^d F}{\partial x_1 \dots \partial x_d} = f(x_1, \dots, x_d).$$

1.4 Convergence Concepts

Let X and X_n, where $n \geqq 1$, denote R^d-valued random variables defined on a probability space $(\Omega, \mathfrak{A}, P)$. The following four convergence concepts find application in probability theory:

a) If there exists a set of measure zero $N \in \mathfrak{A}$ such that, for all $\omega \notin N$, the sequence of the $X_n(\omega) \in R^d$ converges in the usual sense to $X(\omega) \in R^d$, then $\{X_n\}$ is said to converge **almost certainly** $[P]$ or **with probability 1** to X. We write

$$\underset{n \to \infty}{\text{ac-lim}}\ X_n = X.$$

b) If, for every $\varepsilon > 0$,

$$p_n(\varepsilon) = P\{\omega : |X_n(\omega) - X(\omega)| > \varepsilon\} \to 0 \quad (n \to \infty),$$

then $\{X_n\}$ is said to converge **stochastically** or **in probability** to X, and we write

$$\underset{n \to \infty}{\text{st-lim}}\ X_n = X.$$

c) If X_n and X belong to L^p and $E|X_n - X|^p \to 0$, then $\{X_n\}$ is said to converge **in pth mean** to X. For $p = 1$, we speak simply of convergence in mean; for $p = 2$, we speak of convergence **in mean square** or **in quadratic mean** and we write

$$\underset{n \to \infty}{\text{qm-lim}}\ X_n = X.$$

d) Let F_n and F denote the distribution functions of X_n and X. Then, if

$$\lim_{n \to \infty} \int\limits_{R^d} g(x)\, dF_n(x) = \int\limits_{R^d} g(x)\, dF(x)$$

for every real-valued continuous bounded function g defined on R^d, the sequence $\{X_n\}$ is said to converge **in distribution** to X. This is the case if and only if

$$\lim_{n \to \infty} F_n(x) = F(x)$$

at every point at which F is continuous or

$$\lim_{n \to \infty} \varphi_n(t) = \varphi(t) \quad \text{for all } t \in R^d,$$

where φ_n and φ are characteristic functions.

These convergence concepts are in the following relationship to each other:

convergence in qth mean

$\Downarrow$

convergence in pth mean, where $(p \leqq q)$

$\Downarrow$

almost certain convergence $\Rightarrow$ stochastic convergence $\Rightarrow$ convergence in distribution.

Furthermore, a sequence converges stochastically if and only if every subsequence of it contains an almost certainly convergent subsequence. A sufficient condition for $X_n \to X$ (almost certainly) is the condition

$$\sum_{n=1}^{\infty} E |X_n - X|^p < \infty \text{ for some } p > 0.$$

Let $\{X_n\}$ denote a sequence of R^d-valued random variables with distribution $\mathfrak{N}(m_n, C_n)$. This sequence converges in distribution if and only if

$$m_n \rightarrow m, \quad C_n \rightarrow C \quad (n \rightarrow \infty).$$

The limit distribution is $\mathfrak{N}(m, C)$. This follows from consideration of the characteristic functions.

1.5 Products of Probability Spaces, Independence

Suppose that we are given finitely many measurable spaces $(\Omega_i, \mathfrak{A}_i)$ for $i = 1, \ldots, n$. From these spaces, we construct a product measurable space $(\Omega, \mathfrak{A})$ in the following way: the set Ω is the **Cartesian product**

$$\Omega = \underset{i=1}{\overset{n}{\times}} \Omega_i = \Omega_1 \times \ldots \times \Omega_n,$$

consisting of all n-tuples $\omega = (\omega_1, \ldots, \omega_n)$ such that $\omega_i \in \Omega_i$.

In Ω, we define the **product sigma-algebra**

$$\mathfrak{A} = \underset{i=1}{\overset{n}{\times}} \mathfrak{A}_i = \mathfrak{A}_1 \times \ldots \times \mathfrak{A}_n$$

as the sigma-algebra generated by the **cylinder sets** in Ω

$$A = A_1 \times \ldots \times A_n, \quad A_i \in \mathfrak{A}_i.$$

The sigma-algebra $\mathfrak{A}$ is also the smallest sigma-algebra with respect to which the projections $p_i : \Omega \rightarrow \Omega_i$ defined by $p_i(\omega) = \omega_i$ ("projection onto the ith axis") are $(\mathfrak{A} - \mathfrak{A}_i)$-measurable.

If probabilities P_i are given on $(\Omega_i, \mathfrak{A}_i)$, then there exists on $(\Omega, \mathfrak{A})$ exactly one probability P, the so-called **product probability**

$$P = \underset{i=1}{\overset{n}{\times}} P_i = P_1 \times \ldots \times P_n,$$

with the property

$$P(A_1 \times \ldots \times A_n) = P(A_1) \ldots P(A_n) \quad \text{for all } A_i \in \mathfrak{A}_i.$$

This holds also for sigma-finite measures P_i.

In the case $(\Omega_1, \mathfrak{A}_1) = \ldots = (\Omega_n, \mathfrak{A}_n)$, we write for the product space $(\Omega_1^n, \mathfrak{A}_1^n)$. For example, $(R^d, \mathfrak{B}^d) = ((R^1)^d, (\mathfrak{B}^1)^d)$ and Lebesgue measure in R^d is the product of the d one-dimensional Lebesgue measures.

Fubini's theorem, written for the case $n = 2$, is as follows: Let X denote a non-negative or $(P_1 \times P_2)$-integrable $(\mathfrak{A}_1 \times \mathfrak{A}_2 - \mathfrak{B}^1)$-measurable scalar function defined on $\Omega_1 \times \Omega_2$. Then,

$$\int_{\Omega_1 \times \Omega_2} X \, d(P_1 \times P_2) = \int_{\Omega_1} \left(\int_{\Omega_2} X(\omega_1, \omega_2) \, dP_2(\omega_2) \right) dP_1(\omega_1)$$

$$= \int_{\Omega_2} \left(\int_{\Omega_1} X(\omega_1, \omega_2) \, dP_1(\omega_1) \right) dP_2(\omega_2).$$

Product probability spaces serve as models for describing composite experiments that consist in carrying out n individual "mutually unaffected" experiments (statistically independent experiments [see below]) either simultaneously or successively.

The theory of product spaces can be extended to products of an arbitrary family $\{(\Omega_i, \mathfrak{A}_i, P_i)\}_{i \in I}$ (where the index set I is not the empty set) of probability spaces. Suppose that

$$\Omega = \underset{i \in I}{\mathsf{X}} \, \Omega_i$$

is the set of functions ω defined on I and assuming at the point $i \in I$ a value $\omega_i \in \Omega_i$ and that

$$\mathfrak{A} = \underset{i \in I}{\mathsf{X}} \, \mathfrak{A}_i$$

is the sigma algebra in Ω generated by the cylinder sets

$$A = \underset{i \in I}{\mathsf{X}} \, A_i, \quad A_i \in \mathfrak{A}_i, \quad A_i \neq \Omega_i \quad \text{only finitely many times.}$$

Again, $\mathfrak{A}$ is also the smallest sigma-algebra with respect to which every projection p_i defined by $p_i(\omega) = \omega_i$ is measurable. Then, there exists exactly one product probability P defined on $(\Omega, \mathfrak{A})$, that is, one probability P that assigns to the cylinders the value

$$P\left(\underset{i \in I}{\mathsf{X}} \, A_i \right) = \prod_{i \in I} P_i(A_i).$$

The projections p_i have distributions P_i.

A fundamental concept of probability theory is that of independence of events or of random variables. Let $(\Omega, \mathfrak{A}, P)$ denote a probability space. The events $A_1, \dots, A_n$ are said to be **(statistically) independent** if

$$P(A_1 \cap \dots \cap A_n) = P(A_1) \dots P(A_n).$$

Sub-sigma-algebras $\mathfrak{A}_1, \dots, \mathfrak{A}_n$ of $\mathfrak{A}$ are said to be independent if this equation holds for every possible choice of events $A_i \in \mathfrak{A}_i$. Finally, random variables $X_1, \ldots, X_n$ (whose ranges may differ for different values of the subscript) are said to be independent if the sigma-algebras $\mathfrak{A}(X_1), \ldots, \mathfrak{A}(X_n)$ generated by them are independent.

Any family of events (sigma-algebras, random variables) is said to be independent if the events belonging to every finite subfamily of that family are independent in the sense of the definition given above.

In a product probability space, the projections are independent random variables. For a given sequence $F_1, F_2, \ldots$ of d-dimensional distribution functions, we can therefore construct an explicit sequence $X_1, X_2, \ldots$ of independent random variables such that the distribution function of X_i is F_i. We do so by choosing $\Omega = (R^d)^\infty = $ set of all sequences $\omega = (\omega_1, \omega_2, \ldots)$ with elements $\omega_i \in R^d, \mathfrak{A} = (\mathfrak{B}^d)^\infty, P = P_{F_1} \times P_{F_2} \times \ldots$, and $X_i(\omega) = p_i(\omega) = \omega_i$.

Let X_1 denote an R^d-valued and X_2 an R^m-valued random variable. Let $F(x_1, x_2)$ denote their joint distribution function and let $F_1(x_1) = F(x_1, \infty)$ and $F_2(x_2) = F(\infty, x_2)$ denote the marginal distributions of X_1 and X_2, respectively. Then, X_1 and X_2 are independent if and only if, for every $x_1 \in R^d$ and $x_2 \in R^m$,

$$F(x_1, x_2) = F_1(x_1) F_2(x_2)$$

or, in case of existence of densities,

$$f(x_1, x_2) = f_1(x_1) f_2(x_2).$$

This can be carried over in an analogous manner to the case of $n > 2$ random variables.

Let $X_1, \ldots, X_n$ denote independent real-valued integrable random variables. Then, their product is also integrable and

$$E\left(\prod_{i=1}^{n} X_i\right) = \prod_{i=1}^{n} E(X_i).$$

If the X_i belong to L^2, they are uncorrelated and

$$V(X_1 + \ldots + X_n) = V(X_1) + \ldots + V(X_n).$$

If $X = (X_1, X_2)$ has a normal distribution, uncorrelatedness of X_1 and X_2 implies their independence. If the X_i are independent and have distribution $\mathfrak{N}(m_i, C_i)$, then

$$S_n = \sum_{i=1}^{n} X_i$$

has distribution

$$\mathfrak{N}\left(\sum_{i=1}^{n} m_i, \sum_{i=1}^{n} C_i\right).$$

1.6 Limit Theorems

For an arbitrary sequence $\{A_n\}$ of events in a probability space $(\Omega, \mathfrak{A}, P)$, the set

$$A = \{\omega: \omega \in A_n \text{ for infinitely many } n\}$$

is also an event. With regard to its probability, we have the

Borel-Cantelli lemma. If $\sum P(A_n) < \infty$, then $P(A) = 0$. If the sequence $\{A_n\}$ is independent, then, conversely, $\sum P(A_n) = \infty$ implies $P(A) = 1$. In order to give the simplest possible form to the following theorems, we shall consider an independent sequence $\{X_n\}$ of real-valued identically distributed random variables defined on $(\Omega, \mathfrak{A}, P)$, all with the same distribution function F, and we shall study the limiting behavior of the partial sums

$$S_n = X_1 + \ldots + X_n.$$

The strong law of large numbers.

$$\text{ac -}\lim_{n \to \infty} \frac{S_n}{n} = c \quad \text{(finite)}$$

if and only if $E X_1$ exists. If it does, then $c = E X_1$.

Law of the iterated logarithm. Suppose that $V(X_n) = \sigma^2 < \infty$. If we set $E X_n = \alpha$, then, with probability 1,

$$\limsup_{n \to \infty} \frac{S_n - n\alpha}{\sqrt{2 n \log \log n}} = +|\sigma|$$

and

$$\liminf_{n \to \infty} \frac{S_n - n\alpha}{\sqrt{2 n \log \log n}} = -|\sigma|.$$

(Here, "log" denotes the natural logarithm.)

All versions of the **central limit theorem** assert that the sum of a large number of independent random variables has, under quite general conditions, an approximately normal distribution. In our special case of identically distributed summands, it is as follows: Suppose that $0 < V(X_n) = \sigma^2 < \infty$ and that $E(X_n) = \alpha$. Then, $(S_n - \alpha n)/\sigma \sqrt{n}$ tends in distribution to $\mathfrak{N}(0, 1)$; that is, for all $x \in R^1$,

$$\lim_{n \to \infty} P\left[\frac{S_n - \alpha n}{\sigma \sqrt{n}} \leq x\right] = \frac{1}{\sqrt{2\pi}} \int_{-\infty}^{x} e^{-y^2/2} \, dy.$$

1.7 Conditional Expectations and Conditional Probabilities

Let $(\Omega, \mathfrak{A}, P)$ denote a probability space. The elementary **conditional probability** of an event $A \in \mathfrak{A}$ under the condition $B \in \mathfrak{A}$, where $P(B) > 0$, is

$$P(A|B) = \frac{P(A \cap B)}{P(B)}.$$

However, we frequently encounter a whole family of conditions, each of which has probability 0. Then we need the following more general concept of conditional expectation.

Let $X \in L^1(\Omega, \mathfrak{A}, P)$ denote an R^d-valued random variable and let $\mathfrak{C} \subset \mathfrak{A}$ denote a sub-sigma-algebra of $\mathfrak{A}$. The probability space $(\Omega, \mathfrak{C}, P)$ is a coarsening of the original one and X is, in general, no longer $\mathfrak{C}$-measurable. We seek now a $\mathfrak{C}$-measurable coarsening Y of X that assumes, on the average, the same values as X, that is, an integrable random variable Y such that

$$Y \text{ is } \mathfrak{C} \text{ - measurable,}$$

$$\int_C Y \, dP = \int_C X \, dP \quad \text{for all} \quad C \in \mathfrak{C}.$$

According to the Radon-Nikodym theorem, there exists exactly one such Y, almost certainly unique. It is called the **conditional expectation of X under the condition $\mathfrak{C}$**. We write

$$Y = E(X|\mathfrak{C}).$$

As a special case, we consider a sub-sigma-algebra $\mathfrak{C}$ whose elements are arbitrary unions of countably many "atoms" $\{A_n\}$ such that

$$\bigcup_{n=1}^{\infty} A_n = \Omega, \quad A_n \cap A_m = \varnothing \, (n \neq m).$$

The quantity $E(X|\mathfrak{C})$ is constant on the sets A_n. For $P(A_n) > 0$,

$$E(X|\mathfrak{C})(\omega) = E(X|A_n) = \frac{1}{P(A_n)} \int_{A_n} X \, dP \quad \text{for all } \omega \in A_n,$$

with the value for $P(A_n) = 0$ not specified.

Therefore, the conditional expectation is, for fixed X and $\mathfrak{C}$, a function of $\omega \in \Omega$. It follows from the definition that, in particular,

$$E(E(X|\mathfrak{C})) = E(X)$$

and

$$|E(X|\mathfrak{C})| \leq E(|X| \, |\mathfrak{C}) \quad \text{almost certainly.}$$

Other important properties of the conditional expectation are as follows (all the equations and inequalities shown hold with probability 1):

a) $\quad \mathfrak{C} = \{\emptyset, \Omega\} \quad \Rightarrow E(X|\mathfrak{C}) = E(X),$

b) $\quad X \geq 0 \qquad\quad \Rightarrow E(X|\mathfrak{C}) \geq 0,$

c) $\quad X \ \mathfrak{C}\text{-measurable} \Rightarrow E(X|\mathfrak{C}) = X,$

d) $\quad X = \text{const} = a \Rightarrow E(X|\mathfrak{C}) = a,$

e) $\quad X, Y \in L^1 \qquad \Rightarrow E(aX + bY|\mathfrak{C}) = a E(X|\mathfrak{C}) + b E(Y|\mathfrak{C}),$

f) $\quad X \leq Y \qquad\quad \Rightarrow E(X|\mathfrak{C}) \leq E(Y|\mathfrak{C}),$

g) $\quad X \ \mathfrak{C}\text{-measurable}, \ X, X Y' \in L^1 \Rightarrow E(X Y'|\mathfrak{C}) = X E(Y'|\mathfrak{C}),$

$\qquad\qquad$ in particular $E(E(X|\mathfrak{C}) Y'|\mathfrak{C}) = E(X|\mathfrak{C}) E(Y'|\mathfrak{C}),$

h) $\quad X, \mathfrak{C} \text{ independent} \Rightarrow E(X|\mathfrak{C}) = E(X).$

For later use, we point out in particular, that, for $\mathfrak{C}_1 \subset \mathfrak{C}_2 \subset \mathfrak{A}$,

(1.7.1) $\qquad E(E(X|\mathfrak{C}_2)|\mathfrak{C}_1) = E(E(X|\mathfrak{C}_1)|\mathfrak{C}_2) = E(X|\mathfrak{C}_1).$

The **conditional probability** $P(A|\mathfrak{C})$ of an event A under the condition $\mathfrak{C} \subset \mathfrak{A}$ is defined by

$$P(A|\mathfrak{C}) = E(I_A|\mathfrak{C}).$$

Being a conditional expectation, the conditional probability is a $\mathfrak{C}$-measurable function on Ω. In particular, for a $\mathfrak{C}$ generated by at most countably many atoms $\{A_n\}$,

$$P(A|\mathfrak{C})(\omega) = \frac{P(A \cap A_n)}{P(A_n)} \quad \text{for all} \ \ \omega \in A_n \text{ such that } P(A_n) > 0.$$

From the properties of conditional expectation, it follows in particular that $0 \leq P(A|\mathfrak{C}) \leq 1, P(\emptyset|\mathfrak{C}) = 0, P(\Omega|\mathfrak{C}) = 1,$ and

$$P\left(\bigcup_{n=1}^{\infty} A_n \Big| \mathfrak{C}\right) = \sum_{n=1}^{\infty} P(A_n|\mathfrak{C}), \quad \{A_n\} \quad \text{pairwise disjoint in} \quad \mathfrak{A},$$

with probability 1. However, since $P(A|\mathfrak{C})$ is defined only up to a set of measure 0 depending on A, it does *not* follow that $P(\cdot|\mathfrak{C})$ is, for fixed $\omega \in \Omega$, a probability on $\mathfrak{A}$. On the other hand, for a random variable X, consider the conditional probability

$$P(X \in B|\mathfrak{C}) = P(\{\omega: X(\omega) \in B\}|\mathfrak{C}), \quad B \in \mathfrak{B}^d.$$

There exists a function $p(\omega, B)$ defined on $\Omega \times \mathfrak{B}^d$ with the following properties: For fixed $\omega \in \Omega$, the function $p(\omega, \cdot)$ is a probability on $\mathfrak{B}^d$; for fixed B, the function $p(\cdot, B)$ is a version of $P(X \in B|\mathfrak{C})$; that is, $p(\cdot, B)$ is $\mathfrak{C}$-measurable and

$$P(C \cap [X \in B]) = \int_C p(\omega, B)\, dP(\omega) \quad \text{for all } C \in \mathfrak{C}.$$

Such a function p (uniquely defined up to a set of measure 0 in $\mathfrak{C}$ that is independent of B) is called the **conditional (probability) distribution** of X for given $\mathfrak{C}$. For $g(X) \in L^1$,

$$E(g(X)|\mathfrak{C}) = \int_{R^d} g(x)\, p(\omega, dx).$$

If $(\Omega', \mathfrak{A}')$ is a measurable space and $\mathfrak{C} = \mathfrak{A}(Y)$ is the sigma-algebra generated by an Ω'-valued random variable Y, we write

$$E(X|\mathfrak{C}) = E(X|Y).$$

For every $\mathfrak{A}(Y)$-measurable random variable Z, there exists a measurable function h such that $Z = h(Y)$; that is, the value of Z is fixed by the value of Y at ω. This h is uniquely defined up to a set N of images of Y such that $P[Y \in N] = 0$. For the conditional expectation $E(X|Y)$, this means the existence of a measurable function h defined on Ω' such that

$$E(X|Y) = h(Y).$$

We write suggestively

$$h(y) = E(X|Y = y).$$

In the special case of the conditional probability $P(X \in B|Y)$, we go on to the conditional distribution $p(\omega, B)$, for which there is now an almost certainly unique q such that

$$p(\omega, B) = q(Y(\omega), B).$$

We also write

$$q(y, B) = P(X \in B|Y = y),$$

which is measurable with respect to y and, for fixed y, is a probability with respect to B. For $g(X) \in L^1$,

$$(1.7.2) \qquad E(g(X)|Y = y) = \int_{R^d} g(x)\, q(y, dx) = \int_{R^d} g(x)\, P(dx|Y = y)$$

and

$$P[X \in B] = \int_{\Omega'} P(X \in B|Y = y)\, dP_Y(\omega'),$$

where P_Y is the distribution of Y.

If $q(y, \cdot) = P(X \in \cdot |Y = y)$ has a density $h(x, y)$ in R^d, this density is called the **conditional density of** X **under the condition** $Y = y$.

Example. Let X denote an R^d-valued and Y an R^p-valued random variable and suppose that their joint distribution in R^{d+p} has a density $f(x, y)$. Then, $P(X \in B | Y = y)$ has a density $h(x, y)$ which, for all y such that the marginal density

$$f_2(y) = \int_{R^d} f(x, y) \, dx$$

of Y is positive, has the form

$$h(x, y) = \frac{f(x, y)}{f_2(y)}.$$

For all integrable functions $g(x)$, we have, in accordance with formula (1.7.2),

$$E(g(X) | Y = y) = \frac{\int_{R^d} g(x) f(x, y) \, dx}{f_2(y)}.$$

If X and Y are independent, then $f(x, y) = f_1(x) f_2(y)$, so that $h(x, y) = f_1(x)$.

Suppose that the joint distribution of X and Y is a normal distribution $\mathfrak{N}(m, C)$. Let us write $EX = m_x$, $EY = m_y$, and

$$C = \begin{pmatrix} E(X - m_x)(X - m_x)' & E(X - m_x)(Y - m_y)' \\ E(Y - m_y)(X - m_x)' & E(Y - m_y)(Y - m_y)' \end{pmatrix} = \begin{pmatrix} C_x & C_{xy} \\ C_{yx} & C_y \end{pmatrix}.$$

Then, for positive-definite C_y, the conditional density of X given $Y = y$ is the density of the d-dimensional normal distribution

$$\mathfrak{N}(m_x + C_{xy} C_y^{-1}(y - m_y), C_x - C_{xy} C_y^{-1} C_{yx}).$$

1.8 Stochastic Processes

The example cited in section 1.1 of the measurement of water level during an interval of time $[t_0, T]$ and the description of the position of a particle subject to Brownian motion as a function of time make it necessary to consider simultaneously a family of random variables that depend on a continuous parameter (time).

More generally, let I denote an arbitrary nonempty index set and let $(\Omega, \mathfrak{A}, P)$ denote a probability space. Then, a family $\{X_t; t \in I\}$ of R^d-valued random variables is called a **stochastic process (random process, random function)** with **parameter set (index set)** I and state space R^d.

If I is finite, we are dealing simply with finitely many random variables. In the case $I = \{ \ldots, -1, 0, 1, \ldots \}$ or $\{1, 2, \ldots\}$, we speak of a random sequence or a time series. It is preferable to reserve the term "process" for I uncountably infinite.

In what follows, I is always an interval $[t_0, T]$, where $t_0 < T$, of the real axis R^1. We interpret the parameter t as time. We wish to admit the cases $t_0 = -\infty$ and $T = \infty$. Then, $[t_0, T]$ is interpreted as $(-\infty, T], [t_0, \infty)$, or $(-\infty, \infty)$.

If $\{X_t; \, t \in [t_0, T]\}$ is a stochastic process, then, $X_t(\cdot)$ is, for every fixed $t \in [t_0, T]$, an R^d-valued random variable whereas, for every fixed $\omega \in \Omega$ (hence for every observation), $X.(\omega)$ is an R^d-valued function defined on $[t_0, T]$, hence an element of the product space $(R^d)^{[t_0, T]}$. It is called a **sample function (realization, trajectory, path)** of the stochastic process.

The **finite-dimensional distributions** of a stochastic process $\{X_t; \, t \in [t_0, T]\}$ are given by

$$P[X_t \leqq x] = F_t(x),$$

$$P[X_{t_1} \leqq x_1, \, X_{t_2} \leqq x_2] = F_{t_1, t_2}(x_1, x_2),$$

$$\vdots \qquad\qquad\qquad \vdots$$

$$P[X_{t_1} \leqq x_1, \ldots, X_{t_n} \leqq x_n] = F_{t_1, \ldots, t_n}(x_1, \ldots, x_n),$$

$$\vdots \qquad\qquad\qquad \vdots$$

where t and t_i belong to $[t_0, T]$, x and x_i belong to R^d (the symbol $\leqq$ applies to the components), and $n \geqq 1$.

Obviously, this system of distribution functions satisfies the following two conditions:

a) *Condition of symmetry:* if $\{i_1, \ldots, i_n\}$ is a permutation of the numbers 1, $\ldots, n$, then for arbitrary instants and $n \geqq 1$,

$$F_{t_{i_1}, \ldots, t_{i_n}}(x_{i_1}, \ldots, x_{i_n}) = F_{t_1, \ldots, t_n}(x_1, \ldots, x_n).$$

b) *Condition of compatibility:* for $m < n$ and arbitrary $t_{m+1}, \ldots, t_n \in [t_0, T]$,

$$F_{t_1, \ldots, t_m, t_{m+1}, \ldots, t_n}(x_1, \ldots, x_m, \infty, \ldots, \infty) = F_{t_1, \ldots, t_m}(x_1, \ldots, x_m).$$

In many practical cases, we are given not a family of random variables defined on a probability space but a family of distributions $P_{t_1, \ldots, t_n}(B_1, \ldots, B_n)$ or their distribution functions $F_{t_1, \ldots, t_n}(x_1, \ldots, x_n)$ which satisfy the symmetry and compatibility conditions. That these two concepts are equivalent is seen from the following theorem:

(1.8.1) Kolmogorov's fundamental theorem. For every family of distribution functions that satisfy the symmetry and compatibility conditions, there exists a probability space $(\Omega, \mathfrak{A}, P)$ and a stochastic process $\{X_t; t \in [t_0, T]\}$ defined on it that possesses the given distributions as finite-dimensional distributions.

In particular, if we start off with given distributions, we shall always assume the following choice to have been made:

$$\Omega = (R^d)^{[t_0, T]} = \text{set of all } R^d\text{-valued functions } \omega = \omega (\cdot) \text{ defined on } [t_0, T],$$

$$\mathfrak{A} = (\mathfrak{B}^d)^{[t_0, T]} = \text{product sigma-algebra generated by the cylinder sets,}$$

$$X_t (\omega) = \omega (t) = \text{projection of } \omega \text{ onto the "}t\text{-axis", that is, the value of the function } \omega \text{ at the point } t.$$

Now, the probability P on $(\Omega, \mathfrak{A})$ is not (as in section 1.5 for independent X_t) simply the product probability but is determined on the cylinder sets by

$$P \{\omega: \omega (t_1) \in B_1, \ldots, \omega (t_n) \in B_n\} = P_{t_1, \ldots, t_n} (B_1, \ldots, B_n)$$

and can be continued in a unique manner to all $\mathfrak{A}$. Henceforth, this canonical choice, for which the elementary events coincide with the sample functions, will be the one made.

Also, for $\{X_t; t \in [t_0, T]\}$ we shall write briefly X_t or $X(t)$, usually omitting the variable ω.

Two stochastic processes X_t and $\overline{X}_t$ defined on the same probability space are said to be **(stochastically) equivalent** if, for every $t \in [t_0, T]$, we have $X_t = \overline{X}_t$ with probability 1. Then, $\overline{X}_t$ is called a **version** of X_t and vice versa. The finite-dimensional distributions of X_t and $\overline{X}_t$ coincide. However, since the set N_t of exceptional values of ω for which $X_t \neq \overline{X}_t$ depends in general on t, the sample functions of equivalent processes can have quite different analytica properties. For example, for $\Omega = [t_0, T] = [0, 1]$ and $P = \lambda$, the processes $X_t (\omega) \equiv 0$ and

$$\overline{X}_t (\omega) = \begin{cases} 0, & \omega \neq t, \\ 1, & \omega = t, \end{cases}$$

are equivalent. Nonetheless, every sample function of X_t is a continuous function whereas no sample function of $\overline{X}_t$ is. To avoid such phenomena, we shall in what follows always assume without further mention that a separable version (which always exists) has been chosen. A process X_t is said to be **separable** if there exists a countable set $\{t_1, t_2, \ldots\} = M$ of instants, dense in the interval $[t_0, T]$, and a set $N \in \mathfrak{A}$ of P-measure 0 such that, for every open subinterval (a, b) of $[t_0, T]$ and every closed subset A of R^d, the two sets

$$\{\omega: X_t (\omega) \in A \text{ for all } t \in (a, b) \cap M\} \in \mathfrak{A}$$

and

$$\{\omega: X_t(\omega) \in A \text{ for all } t \in (a, b)\} \quad \text{(in general not measurable)}$$

differ, if at all, only on a subset of N. If we arrange for all subsets of the sets of measure zero to belong to $\mathfrak{A}$ (which is always possible), the second set will also belong to $\mathfrak{A}$ and will possess the same probability as the first.

How can we tell from the finite-dimensional distributions of a process whether this process has continuous sample functions or not? The following **criterion of Kolmogorov** asserts that only the two-dimensional distributions are necessary for this: Let a, b, and c denote positive numbers such that, for t and s in $[t_0, T]$,

(1.8.2) $E |X_t - X_s|^a \leqq c |t - s|^{1+b}$.

Then, (a separable version of) X_t possesses with probability 1 continuous sample functions.

A stochastic process is said to be **(strictly) stationary** if its finite-dimensional distributions are invariant under time displacements, that is, if, for t_i, $t_i + t \in [t_0, T]$,

$$F_{t_1+t, \ldots, t_n+t}(x_1, \ldots, x_n) = F_{t_1, \ldots, t_n}(x_1, \ldots, x_n).$$

For stationary processes, t is generally a member of R^1. If in addition $X_t \in L^2$ for all t, it then follows that $E X_t = m = $ const and $\text{Cov}(X_t, X_s) = C(t-s)$. A process with the last two properties is said to be **stationary in the wide sense**. If it has the property $\lim_{t \to s} E |X_t - X_s|^2 = 0$ (mean-square continuity), then the covariance matrix C has the representation

$$C(t) = \int\limits_{-\infty}^{\infty} e^{itu} \, dF(u), \quad -\infty < t < \infty,$$

where the $d \times d$ matrix $F^{\cdot}(u) = (F_{ij}(u))$ is the so-called **spectral distribution function** of X_t. This matrix has the following properties: a) for arbitrary $u_1 < u_2$, the matrix $F(u_2) - F(u_1)$ is nonnegative-definite and b) tr $(F(\infty) - F(-\infty)) < \infty$. From an intuitive standpoint, F gives the distribution of the frequencies of the harmonic oscillations participating in the construction of X_t. If F has a density f, this is called the spectral density of X_t. If

$$\int\limits_{-\infty}^{\infty} |C(t)| \, dt < \infty,$$

f is obtained from the inversion formula

$$f(u) = \frac{1}{2\pi} \int\limits_{-\infty}^{\infty} e^{-itu} C(t) \, dt.$$

An R^d-valued stochastic process is called a **Gaussian process** if its finite-dimensional distributions are normal distributions, hence, if the joint distribution of $X_{t_1}, \dots, X_{t_n}$ has the following characteristic function:

$$\varphi_{t_1, \dots, t_n}(u_1, \dots, u_n) = \exp\left(i \sum_{k=1}^{n} u'_k\, m(t_k) - \frac{1}{2} \sum_{k=1}^{n} \sum_{j=1}^{n} u'_k\, C(t_k, t_j)\, u_j\right),$$

$$u_1, \dots, u_n \in R^d, \quad t_1, \dots, t_n \in [t_0, T].$$

Here, $m(t) = E X_t$ and $C(t, s) = \text{Cov}(X_t, X_s)$.

Therefore, the finite-dimensional distributions of a Gaussian process are uniquely determined by the two functions $m(t)$ and $C(t, s)$ (hence by the first and second moments). A Gaussian process that is stationary in the wide sense is strictly stationary.

1.9 Martingales

Let $(\Omega, \mathfrak{A}, P)$ denote a probability space, let $\{X_t; t \in [t_0, T]\}$ denote an R^d-valued stochastic process defined on $(\Omega, \mathfrak{A}, P)$, and let $\{\mathfrak{A}_t\}_{t \in [t_0, T]}$ denote an increasing family of sub-sigma-algebras of $\mathfrak{A}$, that is, one having the property

$$\mathfrak{A}_s \subset \mathfrak{A}_t \quad \text{for} \quad t_0 \leqq s \leqq t \leqq T.$$

If X_t is $\mathfrak{A}_t$-measurable and integrable for all t, then the pair $\{X_t, \mathfrak{A}_t\}_{t \in [t_0, T]}$ is called a **martingale** if

$$E(X_t \mid \mathfrak{A}_s) = X_s \quad \text{almost certainly}$$

for all s and t in $[t_0, T]$, where $s \leqq t$. If X_t is a real-valued process and if we replace the equality sign in the last formula with $\leqq$ or $\geqq$, what we have is a **supermartingale** or a **submartingale**. In particular, if

$$\mathfrak{A}_t = \mathfrak{A}([t_0, t]) = \mathfrak{A}(X_s; t_0 \leqq s \leqq t),$$

that is, the history of the process X_t prior to the instant t, is chosen as a condition, then X_t is called a martingale (or a supermartingale or submartingale).

Martingales are an abstract presentation of the concept of "fair game" and they constitute one of the most important tools in the theory of stochastic processes. Sample functions of a (separable) martingale have no discontinuities of the second kind; that is, they have, at worst, jumps.

Let X_t and Y_t denote two martingales with respect to the same monotonic family $\mathfrak{A}_t$. Then, $A X_t + B Y_t$ (where A and B are fixed $p \times d$ matrices) is a martingale and, in particular, $X_t - X_{t_0}$ is a martingale. Furthermore, for every martingale X_t, the process $|X_t|^p$ (where $p \geqq 1$) is a submartingale whenever $X_t \in L^p$. For a real-valued martingale X_t, $X_t^+ = \max(X_t, 0)$ and $X_t^- = \max$

$(-X_t, 0)$ are submartingales. The martingale X_t is a submartingale if and only if $-X_t$ is a supermartingale. For a supermartingale (resp. submartingale), $E\,X_t$ is a monotonically decreasing (resp. increasing) function. We have the following

Convergence theorem: If $\{X_t, \mathfrak{A}_t\}$ is a supermartingale that satisfies the condition

$$\sup_{[t_0, T]} E\ X_t^- < \infty,$$

then the limit

$$\text{ac -}\lim_{t \to T} X_t = X$$

exists and belongs to L^1. This holds in particular for $X_t \geqq 0$ or

$$\sup_{[t_0, T]} E\,(X_t) < \infty.$$

If $[a, b]$ is a bounded subinterval of $[t_0, T]$, then the so-called **supermartingale inequalities**

$$c\,P\,[\sup_{[a, b]} X_t \geqq c] \leqq E\,X_a + E\,X_b^-,$$

$$c\,P\,[\inf_{[a, b]} X_t \leqq -c] \leqq E\,X_b^-$$

hold for every supermartingale and positive number c.

Since the process $-|X_t|^p$ is a supermartingale if $X_t \in L^p$ (where $p \geqq 1$) is an R^d-dimensional martingale, it follows in particular that

$$(1.9.1) \qquad P\,[\sup_{[a, b]} |X_t| \geqq c] \leqq E\,|X_b|^p / c^p \quad \text{for all } c > 0.$$

Furthermore, for every martingale $X_t \in L^p$ (where $p > 1$),

$$(1.9.2) \qquad E\,(\sup_{[a, b]} |X_t|^p) \leqq \left(\frac{p}{p-1}\right)^p E\,|X_b|^p.$$

Chapter 2
Markov Processes and Diffusion Processes

2.1 The Markov Property

The groundwork for the theory of Markov stochastic processes was laid in 1906 by A. A. Markov who, in his investigation of connected experiments, formulated the principle (now named after him) that the "future" is independent of the "past" when we know the "present".

On the other hand, this principle is the causality principle of classical physics carried over to stochastic dynamic systems. It specifies that knowledge of the state of a system at a given time is sufficient to determine its state at any future time. An analytical example of this may be seen in the theory of ordinary differential equations: the differential equation

$$\dot{x}_t = f(t, x_t)$$

states that the change taking place in x_t at time t depends only on x_t and t and not on the values of x_s for $s < t$. A consequence of this is that, under certain conditions on f, the solution curve for x_t is uniquely determined by an initial point (t_0, c):

$$x_t = x_t(t_0, c), \quad t > t_0, \quad x_{t_0} = c.$$

Further information about the state x_s at previous times $s < t$ is therefore not necessary for determination of the solution curve. We say that the system has no after-effects or "no memory".

If we carry this idea over to stochastic dynamic systems, we get the Markov property. It says that *if the state of a system at a particular time s (the present) is known, additional information regarding the behavior of the system at times t < s (the past) has no effect on our knowledge of the probable development of the system at t > s (in the future).*

We shall now give a formal mathematical definition of the Markov property as a property of certain stochastic processes.

Let $\{X_t; t \in [t_0, T]\}$ denote a stochastic process whose state space is a d-dimensional Euclidean space R^d (for $d \geq 1$) and whose index set is an interval $[t_0, T]$

of the real axis R^1. For our purposes, it will be sufficient in all cases to assume

$$[t_0, T] \subset [0, \infty) = R^+.$$

Thus, $0 \leq t_0 < \infty$. Here, we admit $T = \infty$, in which case $[t_0, T]$ should be interpreted as $[t_0, \infty)$. For $\{X_t; \ t \in [t_0, T]\}$ we shall write simply X_t. We shall refer to the index t as the "time". We shall always assume that the state space R^d is endowed with the sigma-algebra $\mathfrak{B}^d$ of Borel sets.

The process X_t is defined on a certain probability space $(\Omega, \mathfrak{A}, P)$. A sample function $X.(\omega)$ is therefore an R^d-valued function defined on the interval $[t_0, T]$. We emphasize again that we always assume that the choice

$$\Omega = (R^d)^{[t_0, T]}$$

is made, where $(R^d)^{[t_0, T]}$ is the space of all R^d-valued functions defined on the interval $[t_0, T]$,

$$\mathfrak{A} = (\mathfrak{B}^d)^{[t_0, T]}$$

is the product sigma-algebra generated by the Borel sets in R^d, and $X_t = \omega(t)$ for all $\omega \in \Omega$. Then, P is the probability uniquely defined (according to Kolmogorov's fundamental Theorem (1.8.1)) by the finite-dimensional distributions of the process X_t on $(\Omega, \mathfrak{A})$. If we have further information regarding the analytical properties of the sample functions, we can choose for Ω certain subspaces of $(R^d)^{[t_0, T]}$ (for example, the space of all continuous functions).

Suppose that, for $t_0 \leq t_1 \leq t_2 \leq T$,

$$\mathfrak{A}([t_1, t_2]) = \mathfrak{A}(X_t, \ t_1 \leq t \leq t_2)$$

is the smallest sub-sigma-algebra of $\mathfrak{A}$ with respect to which all the random variables X_t, for $t_1 \leq t \leq t_2$, are measurable. In terms of our time figure, $\mathfrak{A}([t_1, t_2])$ contains the "history" of the process A_t from time t_1 to time t_2, that is, those events that are determined by the conditions imposed on the course of the process X_t during the interval $[t_1, t_2]$ and at no other time. $\mathfrak{A}([t_1, t_2])$ is generated by the cylinder events

$$\{\omega: X_{s_1}(\omega) \in B_1, \ldots, X_{s_n}(\omega) \in B_n\} = [X_{s_1} \in B_1, \ldots, X_{s_n} \in B_n],$$
$$t_1 \leq s_1 < \ldots < s_n \leq t_2, \quad B_i \in \mathfrak{B}^d.$$

We can now formally define a Markov process:

(2.1.1) **Definition.** A stochastic process $\{X_t, t \in [t_0, T]\}$ defined on the probability space $(\Omega, \mathfrak{A}, P)$ with index set $[t_0, T] \subset [0, \infty)$ and with state space R^d is called a **Markov process** (or an elementary or weak Markov process) if the following so-called **Markov property** (or elementary or weak Markov property) is satisfied: for $t_0 \leq s \leq t \leq T$ and all $B \in \mathfrak{B}^d$, the equation

(2.1.2) $P(X_t \in B | \mathfrak{A}([t_0, s])) = P(X_t \in B | X_s)$

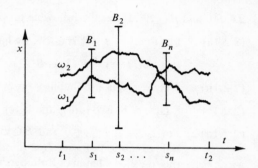

Fig. 1:
Cylinder event.

holds with probability 1. We summarize the various equivalent clarifying formulations of the Markov property in

(2.1.3)   Theorem.   Each of the following conditions is equivalent to the Markov property:

a)   For $t_0 \leqq s \leqq t \leqq T$ and $A \in \mathfrak{A}([t, T])$,

$$P(A \mid \mathfrak{A}([t_0, s])) = P(A \mid X_s).$$

b)   For $t_0 \leqq s \leqq t \leqq T$ and $Y \, \mathfrak{A}([t, T])$-measurable and integrable,

$$E(Y \mid \mathfrak{A}([t_0, s])) = E(Y \mid X_s).$$

c)   For $t_0 \leqq t_1 \leqq t \leqq t_2 \leqq T$, $A_1 \in \mathfrak{A}([t_0, t_1])$ and $A_2 \in \mathfrak{A}([t_2, T])$,

$$P(A_1 \cap A_2 \mid X_t) = P(A_1 \mid X_t) \, P(A_2 \mid X_t).$$

d)   For $n \geqq 1, t_0 \leqq t_1 < \ldots < t_n < t \leqq T$ and $B \in \mathfrak{B}^d$,

$$P(X_t \in B \mid X_{t_1}, \ldots, X_{t_n}) = P(X_t \in B \mid X_{t_n}).$$

(All equations asserting equality of conditional probabilities hold with probability 1).

Proof of these assertions can be found, for example, in Doob [3], pp. 80–85.

(2.1.4)   Remark.   A verbal formulation of Theorem (2.1.3c) is as follows: *Given a Markov process, the past and future are statistically independent when the present is known.*

2.2   Transition Probabilities, the Chapman-Kolmogorov Equation

Let X_t, for $t_0 \leqq t \leqq T$, denote a Markov process. In accordance with what was said in section 1.7, there exists a conditional distribution $q(X_s, B) = P(s, X_s, t, B)$ corresponding to the conditional probability $P(X_t \in B \mid X_s)$. The function $P(s, x, t, B)$ is a function of the four arguments $s, t \in [t_0, T]$ (with $s \leqq t$),

$x \in R^d$, and $B \in \mathfrak{B}^d$. It has the following properties:

(2.2.1a)   For fixed $s \leqq t$ and $B \in \mathfrak{B}^d$, we have with probability 1

$$P(s, X_s, t, B) = P(X_t \in B | X_s).$$

(2.2.1b) $P(s, x, t, \cdot)$ is a probability on $\mathfrak{B}^d$ for fixed $s \leqq t$ and $x \in R^d$.

(2.2.1c) $P(s, \cdot, t, B)$ is $\mathfrak{B}^d$-measurable for fixed $s \leqq t$ and $B \in \mathfrak{B}^d$.

(2.2.1d)   For $t_0 \leqq s \leqq u \leqq t \leqq T$ and $B \in \mathfrak{B}^d$ and for all $x \in R^d$ with the possible exception of a set $N \subset R^d$ such that $P[X_s \in N] = 0$, we have the so-called **Chapman-Kolmogorov equation.**

(2.2.2) $$P(s, x, t, B) = \int_{R^d} P(u, y, t, B)\, P(s, x, u, dy).$$

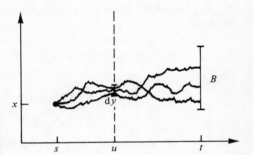

Fig. 2:
The Chapman-Kolmogorov equation.

This can be explained as follows: we have, with probability 1,

$$
\begin{aligned}
P(s, X_s, t, B) &= P(X_t \in B | \mathfrak{A}([t_0, s])) \\
&= E(P(X_t \in B | \mathfrak{A}([t_0, u])) | \mathfrak{A}([t_0, s])) \\
&= E(P(X_t \in B | X_u) | \mathfrak{A}([t_0, s])) \\
&= E(P(u, X_u, t, B) | X_s) \\
&= \int_{R^d} P(u, y, t, B)\, P(s, X_s, u, dy).
\end{aligned}
$$

Here, we used first the Markov property, then the relationship

$$\mathfrak{A}([t_0, s]) \subset \mathfrak{A}([t_0, u]), \quad s \leqq u,$$

and Eq. (1.7.1), again the Markov property, and finally Theorem (2.1.3b) and Eq. (1.7.2).

One can modify the function $P(s, x, t, B)$ in such a way that (2.2.2) holds for *all* $x \in R^d$ (without destroying properties (a)-(c)). Henceforth, we shall always assume this to have been done.

Furthermore, it is always possible to choose $P(s, x, t, B)$ in such a way that

(2.2.1e)   For all $s \in [t_0, T]$ and $B \in \mathfrak{B}^d$, we have

$$P(s, x, s, B) = I_B(x) = \begin{cases} 1 & \text{for} \quad x \in B, \\ 0 & \text{for} \quad x \notin B. \end{cases}$$

This last statement follows from the fact that

$$P(X_s \in B | X_s) = I_{[X_s \in B]}$$

for all $[P]$ values $X_s = x$.

(2.2.3) **Definition.** A function $P(s, x, t, B)$ with the properties (2.2.1b–e) (where (2.2.2) is satisfied for all $x \in R^d$) is called a **transition probability (transition function)**. If X_t is a Markov process and $P(s, x, t, B)$ is a transition probability, so that (2.2.1a) is satisfied, then $P(s, x, t, B)$ is called a transition probability of the Markov process X_t. Then, for fixed $s, t \in [t_0, T]$ such that $s \leq t$, it is uniquely defined as a function of x and B with the possible exception of a set N of values of x (independent of B) such that $P[X_s \in N] = 0$.

We shall also use the notation

$$P(s, x, t, B) = P(X_t \in B | X_s = x),$$

which is the probability that the observed process will be in the set B at time t if at time s, where $s \leq t$, it was in the state x. Here, the number $P(X_t \in B | X_s = x)$ is completely defined by the equation above, even though the condition $[X_s = x]$ may have probability 0 (as it does for most of the processes examined in the present book).

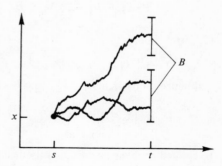

Fig. 3:
The transition probability
$P(s, x, t, B)$.

(2.2.4) **Remarks.** (a) If the probability $P(s, x, t, \cdot)$ has a density, that is, if, for all $s, t \in [t_0, T]$, where $s < t$ (for $s = t$, existence of a density is impossible by virtue of (2.2.1e)), all $x \in R^d$, and all $B \in \mathfrak{B}^d$, we have

$$P(s, x, t, B) = \int_B p(s, x, t, y)\, dy,$$

where $p(s, x, t, y)$ is a nonnegative-valued function that is measurable with respect to y and whose integral is equal to 1, then Eq. (2.2.2) reduces to

$$p(s, x, t, y) = \int_{R^d} p(s, x, u, z) \, p(u, z, t, y) \, dz.$$

In imprecise language, this means that the probability of a transition from x at time s to y at time t is equal to the probability of the transition to z at an intermediate time u, multiplied by the probability of the transition from z at the time u to y at the time t, summed over all intermediate values z.

b) In (2.2.1e), we can replace $I_B(x)$ with $\delta_x(B)$ (where δ_x is the probability measure concentrated at the point x). What (2.2.1e) says is that the state of the process does not change during a time interval of length 0.

The significance of the transition probabilities for Markov processes is that all finite-dimensional distributions of the process can be obtained from them and from the initial distribution at time t_0,. More precisely (see, for example, Krickeberg [7], p. 151), we have

(2.2.5) **Theorem.** If X_t is a Markov process in $[t_0, T]$, if $P(s, x, t, B)$ is its transition probability, and if P_{t_0} is the distribution of X_{t_0}, that is,

$$P_{t_0}(A) = P[X_{t_0} \in A],$$

then, for finite-dimensional distributions

$$P[X_{t_1} \in B_1, \dots, X_{t_n} \in B_n], \quad t_0 \leqq t_1 < \dots < t_n \leqq T, \quad B_i \in \mathfrak{B}^d,$$

we have

$$P[X_{t_1} \in B_1, \dots, X_{t_n} \in B_n] = \int_{R^d} \int_{B_1} \dots \int_{B_{n-1}} P(t_{n-1}, x_{n-1}, t_n, B_n) \cdot$$

(2.2.6)

$$\cdot P(t_{n-2}, x_{n-2}, t_{n-1}, dx_{n-1}) \dots P(t_0, x_0, t_1, dx_1) \, P_{t_0}(dx_0),$$

and hence, in particular,

$$P[X_t \in B] = \int_{R^d} P(t_0, x, t, B) \, P_{t_0}(dx).$$

In applied problems we frequently deal with transition probabilities in the sense of the definition (2.2.3), rather than Markov processes in the sense of the definition (2.1.1), and we must first construct the process as a family of random variables. That this is always possible is asserted by the following theorem, which also gives us a second and more convenient way of defining a Markov process.

(2.2.7) **Theorem.** Let $P(s, x, t, B)$ denote a transition probability, where s, $t \in [t_0, T]$. Then for every initial probability P_{t_0} on $\mathfrak{B}^d$ there exist a probability space $(\Omega, \mathfrak{A}, P)$ and a Markov process X_t (where $t \in [t_0, T]$) defined on it, which has transition probability $P(s, x, t, B)$ and for which X_{t_0} has the distribution P_{t_0}.

To prove this, we use equation (2.2.6) to construct from $P(s, x, t, B)$ and P_{t_0} consistent finite-dimensional distributions and from them, in accordance with Kolmogorov's fundamental theorem (1.8.1), the desired process. Here, we can always use for Ω, X_t, and $\mathfrak{A}$ the special choice discussed in section 2.1.

(2.2.8) Definition. A Markov process X_t for $t \in [t_0, T]$ is said to be homogeneous (with respect to time) if its transition probability $P(s, x, t, B)$ is stationary, that is, if the condition

$$P(s+u, x, t+u, B) = P(s, x, t, B)$$

is identically satisfied for $t_0 \leqq s \leqq t \leqq T$ and $t_0 \leqq s+u \leqq t+u \leqq T$. In this case, the transition probability is thus a function only of $x, t-s$, and B. Hence, we can write it in the form

$$P(t-s, x, B) = P(s, x, t, B), \quad 0 \leqq t-s \leqq T - t_0.$$

Therefore, $P(t, x, B)$ is the probability of transition from x to B in time t, regardless of the actual position of the interval of length t on the time axis. For homogeneous processes, the Chapman-Kolmogorov equation becomes

$$P(t+s, x, B) = \int_{R^d} P(s, y, B) P(t, x, dy).$$

As a rule, homogeneous Markov processes are defined on an interval of the form $[t_0, \infty)$, so that the transition probability $P(t, x, B)$ is defined for $t \in [0, \infty)$.

(2.2.9) Remark. Every Markov process X_t can, by assuming time to be a state component, be transformed into a homogeneous Markov process $Y_t = (t, X_t)$ with state space $[t_0, T] \times R^d$. The transition probability $Q(t, y, B)$ of Y_t for the special sets $B = C \times D$, $C \in \mathfrak{B}^1([t_0, T])$, $D \in \mathfrak{B}^d$ is then given by

$$(2.2.10) \quad Q(t, y, C \times D) = Q(t, (s, x), C \times D) = P(s, x, s+t, D) I_C(s+t).$$

This uniquely determines the probability $Q(t, y, \cdot)$ on the entire set $\mathfrak{B}^1([t_0, T]) \times \mathfrak{B}^d$.

(2.2.11) Remark. When is the Markov process X_t, for $t \in [t_0, T]$, also a stationary process? A necessary and sufficient condition for stationarity (see, for example, Khas'minskiy [65], p. 97) is:

a) X_t is homogeneous;
b) there exists an invariant distribution P^0 in the state space, that is,

$$P^0(B) = \int_{R^d} P(t, x, B) P^0(dx) \text{ for all } B \in \mathfrak{B}^d, \ t \in [0, T - t_0].$$

If we choose this P^0 as initial distribution for X_{t_0}, then X_t is a stationary process.

Furthermore, if there exists an invariant P^0, we have, for arbitrary initial distributions and $T = \infty$,

$$\lim_{t \to \infty} P[X_t \in B] = P^0(B)$$

for all $B \in \mathfrak{B}^d$ whose boundary has P^0-measure 0; that is, the invariant distribution is a stationary limit distribution and is in fact independent of the initial distribution. There are probabilistic and analytical conditions under which a homogeneous transition function $P(t, x, B)$ admits an invariant distribution (see Prohorov and Rozanov [15], p. 272, or Khas'minskiy [65], p. 99). Compare Theorem (8.2.12) and remark (9.2.14).

2.3 Examples

By Theorem (2.2.7) an initial and a transition probability fix a Markov process. In the following examples, we shall assume these probabilities given.

(2.3.1) **Example:** *Deterministic motion.* Suppose that to every pair (s, t), where $t_0 \leqq s \leqq t \leqq T$, is assigned a measurable mapping $G_{s,t}$ of R^d into itself such that, for all $x \in R^d$,

$$G_{t,t}(x) = x$$

and

(2.3.2) $G_{u,t}(G_{s,u}(x)) = G_{s,t}(x), \quad s \leqq u \leqq t.$

These equations define in the state space R^d a deterministic motion which shifts into $y = G_{s,t}(x)$, over a time interval of length $t - s$, a point that is at x at time s. A special case is

$$G_{s,t}(x) = x + v(t - s),$$

which describes a uniform motion with velocity $v \in R^d$, or, more generally,

$$G_{s,t}(x) = x_t(s, x),$$

where x_t is the solution of the differential equation

$$\dot{x}_t = f(t, x_t)$$

with initial condition $x_s = x$ (and f is such that there exists a unique solution on the interval $[s, T]$). The corresponding transition probability is

$$P(s, x, t, B) = \delta_{G_{s,t}(x)}(B) = \begin{cases} 1 & \text{for } G_{s,t}(x) \in B, \\ 0 & \text{for } G_{s,t}(x) \notin B. \end{cases}$$

Property (2.2.1b) is obvious, (2.2.1c) follows from the measurability of the mapping $G_{s,t}$, (2.2.1e) follows from the property $G_{s,s}(x) = x$, and the Chapman-

Kolmogorov equation (2.2.1d) follows from (2.3.2) since

$$\int_{R^d} \delta_{G_{u,t}(y)}(B)\, \delta_{G_{s,u}(x)}(dy) = \delta_{G_{u,t}(G_{s,u}(x))}(B).$$

A nontrivial stochastic effect can be achieved only by the choice of the initial probability.

(2.3.3) **Example:** Whether a process is Markovian depends essentially on the choice of the state space. Whereas the solution x_t of the first-order differential equation

$$\dot{x}_t = f(t, x_t), \quad x_{t_0} = c, \quad t_0 \leq t \leq T,$$

is a Markov process (and this also holds for a differential equation of the form

$$\dot{x}_t = f(t, x_t, \xi_t),$$

where ξ_t is a family of independent random variables which are also independent of x_{t_0}), this is not in general true for the solution of an nth-order differential equation

$$x_t^{(n)} = f(t, x_t, \dot{x}_t, \dots, x_t^{(n-1)}).$$

Nevertheless, the customary shift to a first-order differential equation for the $d\,n$-dimensional process

$$y_t = \begin{pmatrix} x_t \\ \dot{x}_t \\ \vdots \\ x_t^{(n-1)} \end{pmatrix}$$

shows that the Markov property holds for y_t.

(2.3.4) **Example:** *Wiener process.* The **Wiener process** is a homogeneous d-dimensional Markov process W_t defined on $[0, \infty)$ with stationary transition probability

$$P(t, x, \cdot) = \begin{cases} \mathfrak{R}(x, t\,I), & t > 0, \\ \delta_x(\cdot), & t = 0; \end{cases}$$

that is, for $t > 0$,

$$P(t, x, B) = P(W_{t+s} \in B \mid W_s = x) = \int_B (2\pi t)^{-d/2}\, e^{-|y-x|^2/2t}\, dy.$$

By virtue of the familiar formula

$$\int_{R^d} n\,(s, x, z)\, n\,(t, z, y)\, \mathrm{d}z = n\,(s + t, x, y),$$

the Chapman-Kolmogorov equation holds for the densities

$$p\,(t, x, y) = n\,(t, x, y) = (2\,\pi\,t)^{-d/2}\, e^{-|y-x|^2/2\,t}\,.$$

As a rule, the initial probability P_0 is taken equal to δ_0; that is, $W_0 = 0$. Since

$$n\,(t, x+z, y+z) = n\,(t, x, y) \text{ for all } z \in R^d,$$

we are dealing with a spacewise as well as a timewise homogeneous process. We shall examine it in greater detail in section 3.1. With the criterion (1.8.2), we shall be able later to show easily that W_t can be chosen in such a way that it possesses *continuous sample functions* with probability 1. Henceforth, we shall assume that W_t was so chosen. The function W_t is a mathematical model of the Brownian motion of a free particle in the absence of friction.

2.4   The Infinitesimal Operator

Just as in example (2.3.1) we assigned to a differential equation $\dot{x}_t = f\,(t, x_t)$ the family $G_{s,t}$ of transformations in the state space, where $G_{s,t}\,(x) = x_t\,(s, x)$ is the value of the solution, at time t, on a solution trajectory that begins at x at time s, we can also assign to a general Markov process X_t a family of mappings, namely operators defined on a function space.

We begin with the discussion of the *homogeneous* case. Let X_t, for $t \in [t_0, T]$, denote a homogeneous Markov process with transition probability $P\,(t, x, B)$. We define the operator T_t on the space $B\,(R^d)$ of bounded measurable scalar functions defined on R^d and equipped with the norm

$$\|g\| = \sup_{x \in R^d} |g\,(x)|$$

as follows: for $t \in [0, T - t_0]$, let $T_t\, g$ denote the function defined by

$$(2.4.1) \qquad T_t\, g\,(x) = E_x\, g\,(X_t) = \int_{R^d} g\,(y)\, P\,(t, x, \mathrm{d}y).$$

Obviously, $T_t\, g\,(x)$ is the mean value (independent of s) of $g\,(X_{s+t})$ under the condition $X_s = x$. Since

$$T_t\, I_B\,(x) = P\,(t, x, B),$$

we can derive the transition probability from the operators T_t. These operators have the following properties:

(2.4.2) **Theorem.**   The operators T_t, for $t \in [0, T - t_0]$ map the space $B\,(R^d)$

into itself, they are linear, positive, and continuous, and they have the norm $\|T_t\| = 1$. The operator T_0 is the identity operator, and

$$T_{s+t} = T_s T_t = T_t T_s, \quad t, s, t+s \in [0, T - t_0].$$

In particular, in the case $T = \infty$ the T_t constitute a commutative one-parameter semigroup, the so-called **semigroup of Markov transition operators**.

(2.4.3) **Examples.** In the case of deterministic motion generated by an autonomous ordinary differential equation $\dot{x}_t = f(x_t)$ we have

$$T_t g(x) = g(x_t(0, x)),$$

where $x_t(0, x)$ is the solution with the initial value $x_0 = x$. For the Wiener process W_t, we have

$$T_t g(x) = (2\pi t)^{-d/2} \int_{R^d} g(y) \, e^{-|y-x|^2/2t} \, dy$$

$$= (2\pi)^{-d/2} \int_{R^d} e^{-|z|^2/2} g(x + \sqrt{t}\, z) \, dz, \quad t > 0.$$

The dynamics of a Markov process may be described by a single operator representing the derivative of the family T_t at the point $t = 0$.

(2.4.4) **Definition.** The **infinitesimal operator (generator)** A of a homogeneous Markov process X_t, for $t_0 \leqq t \leqq T$, is defined by

$$(2.4.5) \qquad A g(x) = \lim_{t \downarrow 0} \frac{T_t g(x) - g(x)}{t}, \quad g \in B(R^d),$$

where the limit is uniform with respect to x (that is, the limit in $B(R^d)$). The domain of definition $D_A \subset B(R^d)$ consists of all functions for which the limit in (2.4.5) exists. The quantity $A g(x)$ is interpreted as the mean infinitesimal rate of change of $g(X_s)$ in case $X_s = x$.

The operator A is in general an unbounded closed linear operator. If the transition probabilities of X_t are stochastically continuous, that is, if, for every $x \in R^d$ and every $\varepsilon > 0$,

$$\lim_{t \downarrow 0} P(t, x, U_\varepsilon) = 1, \quad U_\varepsilon = \{y : |y - x| < \varepsilon\},$$

then $P(t, x, B)$ is uniquely defined by A. In particular, Markov processes with sample functions that are continuous from the right (or, in particular, continuous) have stochastically continuous transition probabilities.

(2.4.6) **Example.** For uniform motion with velocity $v \in R^d$, we have

$$T_t g(x) = g(x + v t)$$

and

$$A g (x) = \lim_{t \downarrow 0} \frac{g (x + v t) - g (x)}{t}$$

$$= \sum_{i=1}^{d} v_i \frac{\partial g (x)}{\partial x_i}.$$

The existence and the required uniformity of the limit are guaranteed in the domain D_A = set of all bounded uniformly continuous functions with bounded uniformly continuous first partial derivatives.

(2.4.7) **Example.** For the d-dimensional Wiener process W_t, we must calculate

$$A g (x) = (2 \pi)^{-d/2} \lim_{t \downarrow 0} \int_{R^d} e^{-|z|^2/2} (g (x + \sqrt{t} z) - g (x)) \, dz$$

in accordance with (2.4.3). For this we use Taylor's theorem, which, for every twice continuously partially differentiable function g, yields

$$g (x + \sqrt{t} z) - g (x) = \sqrt{t} \sum_{i=1}^{d} z_i g_{x_i} (x) + \frac{t}{2} \sum_{i=1}^{d} \sum_{j=1}^{d} z_i z_j g_{x_i x_j} (x)$$

$$+ \frac{t}{2} \sum_{i=1}^{d} \sum_{j=1}^{d} z_i z_j (g_{x_i x_j} (\bar{x}) - g_{x_i x_j} (x)),$$

where $\bar{x}$ is a point between x and $x + \sqrt{t} z$. When we substitute this into the expression given above for $A g (x)$, we get

$$A g = \frac{1}{2} \sum_{i=1}^{d} \frac{\partial^2 g}{\partial x_i^2} = \frac{1}{2} \Delta g,$$

where Δ is the Laplacian operator. The existence and uniformity of the limit are ensured for all g in D_A, which now is the set of bounded twice continuously partially differentiable functions with bounded and uniformly continuous second partial derivatives.

Let us turn now to the *nonhomogeneous* case. Let X_t, for $t \in [t_0, T]$, denote an arbitrary Markov process with transition probability $P (s, x, t, B)$. We refer to the remark (2.2.9), according to which $Y_t = (t, X_t)$ is a homogeneous Markov process with state space $[t_0, T] \times R^d \subset R^{d+1}$. We now define the Markov transition operators T_t and the infinitesimal operator A of X_t as being equal to the same quantities as in the case of the corresponding homogeneous process $Y_t = (t, X_t)$ under the definitions given earlier.

By virtue of (2.2.10),

$$T_t g (s, x) = E_{s,x} g (s+t, X_{t+s}) = \int_{R^d} g (s+t, y) P (s, x, t+s, dy),$$

$$0 \leq t \leq T - s,$$

where $g(s, x)$ is a bounded measurable function in $[t_0, T] \times R^d$ and

$$(2.4.8) \qquad A g(s, x) = \lim_{t \downarrow 0} \frac{T_t g(s, x) - g(s, x)}{t},$$

where the limit means the uniform limit in $(s, x) \in [t_0, T] \times R^d$. Once again, stochastically continuous transition probabilities $P(s, x, t, B)$ (in particular, the transition probabilities of Markov processes with sample functions that are continuous from the right) are uniquely defined by A.

Frequently, it is sufficient to determine the actions of A on only those functions g that are independent of s, that is, to consider

$$(2.4.9) \qquad A g(s, x) = \lim_{t \downarrow 0} \frac{T_t g(s, x) - g(x)}{t}, \quad g \in B(R^d).$$

For homogeneous processes, this reduces to (2.4.5) but, in general, $T_t g$ is, like $A g$, a function of s and x.

2.5 Diffusion Processes

Diffusion processes are special cases of Markov processes with continuous sample functions which serve as probability-theoretic models of physical diffusion phenomena. The simplest and oldest example is the motion of very small particles, such as grains of pollen in a fluid, the so-called Brownian motion. The Wiener process W_t of example (2.3.4) is a mathematical model of this timewise homogeneous phenomenon in a homogeneous medium (see section 12.1 and Wax [51]).

Besides the original significance of the diffusion process, there is another one, which is emphasized in this book, namely, the description of technical systems subject to "white noise". Also, continuous models for random-walk problems lead to diffusion processes.

Depending on the classification of methods (see the Introduction), there are two basically different approaches to the class of diffusion processes. On the one hand, one can define them in terms of the conditions on the transition probabilities $P(s, x, t, B)$, which is what we shall do in the present section. On the other hand, one can study the state X_t itself and its variation with respect to time. This leads to a stochastic differential equation for X_t. As we shall see in Chapter 9, the two approaches lead essentially to the same class of processes.

(2.5.1) **Definition.** A Markov process X_t, for $t_0 \leq t \leq T$, with values in R^d and almost certainly continuous sample functions is called a **diffusion process** if its transition probability $P(s, x, t, B)$ satisfies the following three conditions for every $s \in [t_0, T)$, $x \in R^d$, and $\varepsilon > 0$:

a) $$\lim_{t \downarrow s} \frac{1}{t-s} \int_{|y-x|>\varepsilon} P(s, x, t, \mathrm{d}y) = 0;$$

b) there exists an R^d-valued function $f(s, x)$ such that

$$\lim_{t \downarrow s} \frac{1}{t-s} \int_{|y-x|\leq\varepsilon} (y-x) P(s, x, t, \mathrm{d}y) = f(s, x);$$

c) there exists a $d \times d$ matrix-valued function $B(s, x)$ such that

$$\lim_{t \downarrow s} \frac{1}{t-s} \int_{|y-x|\leq\varepsilon} (y-x)(y-x)' P(s, x, t, \mathrm{d}y) = B(s, x).$$

The functions f and B are called the **coefficients** of the diffusion process. In particular, f is called the **drift vector** and B is called the **diffusion matrix**. $B(s, x)$, is symmetric and nonnegative-definite.

(2.5.2) Remark. In conditions a) and b) of definition (2.5.1), we had to use truncated moments since $E_{s,x} X_t$ and $E_{s,x} X_t X_t'$ did not necessarily exist. Nonetheless, if, for some $\delta > 0$,

$$\lim_{t \downarrow s} \frac{1}{t-s} E_{s,x} |X_t - X_s|^{2+\delta} = \lim_{t \downarrow s} \frac{1}{t-s} \int_{R^d} |y-x|^{2+\delta} P(s, x, t, \mathrm{d}y) = 0$$

then, since

$$\int_{|y-x|>\varepsilon} |y-x|^k P(s, x, t, \mathrm{d}y) \leq \frac{1}{\varepsilon^{2+\delta-k}} \int_{R^d} |y-x|^{2+\delta} P(s, x, t, \mathrm{d}y)$$

for $k = 0, 1, 2$, condition a) is automatically satisfied and we can choose R^d as region of integration in conditions b) and c).

(2.5.3) Remark. Let us make clear just what conditions a), b), and c) in definition (2.5.1) mean. Condition a) means that large changes in X_t over a short period of time are improbable:

$$P(|X_t - X_s| \leq \varepsilon | X_s = x) = 1 - o(t-s).$$

Let us suppose that the truncated moments in b) and c) are replaced with the usual ones. Then, for the first two moments of the increment $X_t - X_s$ under the condition $X_s = x$ as $t \downarrow s$,

$$E_{s,x}(X_t - X_s) = f(s, x)(t-s) + o(t-s)$$

and

$$E_{s,x}(X_t - X_s)(X_t - X_s)' = B(s, x)(t-s) + o(t-s).$$

Therefore,

$$\text{Cov}_{s,x}(X_t - X_s) = B(s, x)(t - s) + o(t - s),$$

where $\text{Cov}_{s,x}(X_t - X_s)$ is the covariance matrix of $X_t - X_s$ with respect to the probability $P(s, x, t, \cdot)$. Therefore, $f(s, x)$ is the mean velocity vector of the random motion described by X_t under the assumption $X_s = x$, whereas $B(s, x)$ is a measure of the local magnitude of the fluctuation of $X_t - X_s$ about the mean value. If we neglect the last term $o(t - s)$, we can write

$$X_t - X_s \approx f(s, X_s)(t - s) + G(s, X_s)\,\xi,$$

where $E_{s,x}\,\xi = 0$, $\text{Cov}_{s,x}\,\xi = (t - s)\,I$, and $G(s, x)$ is any $d \times d$ matrix with the property $G\,G' = B$. Now, the increments $W_t - W_s$ of the Wiener process have distribution $\mathfrak{N}(0, (t - s)\,I)$. Since we are now only concerned with the distributions, we can write

$$X_t - X_s \approx f(s, X_s)(t - s) + G(s, X_t)(W_t - W_s).$$

The shift (which is usual in analysis) to differentials yields

$$dX_t = f(t, X_t)\,dt + G(t, X_t)\,dW_t.$$

This is a stochastic differential equation that, under a suitable definition of "solution", has as its solution the diffusion process X_t that we started with.

(2.5.4) **Examples.** a) Uniform motion with velocity v is a diffusion process with $f \equiv v$ and $B \equiv 0$.

b) The Wiener process W_t is a diffusion process with drift vector $f \equiv 0$ and diffusion matrix $B \equiv I$.

2.6 Backward and Forward Equations

The decisive property of diffusion processes is that their transition probability $P(s, x, t, B)$ is, under certain regularity assumptions, uniquely determined merely by the drift vector and the diffusion matrix. This is surprising inasmuch as, on the basis of definition (2.5.1), f and B are obtained only from the first two moments of $P(s, x, t, B)$, which do not in general define a distribution.

To each diffusion process with coefficients f and $B = (b_{ij})$ is assigned the second-order differential operator

$$(2.6.1) \qquad \mathfrak{D} \equiv \sum_{i=1}^{d} f_i(s, x)\frac{\partial}{\partial x_i} + \frac{1}{2}\sum_{i=1}^{d}\sum_{j=1}^{d} b_{ij}(s, x)\frac{\partial^2}{\partial x_i\,\partial x_j}.$$

$\mathfrak{D}\,g$ can be formally written for every twice partially differentiable function $g(x)$ and is determined by f and B. In section 2.4, we saw that every diffusion process is uniquely determined by its infinitesimal operator A. We calculate this operator from

(2.6.2) $A g (s, x) = \lim\limits_{t \downarrow 0} \dfrac{1}{t} \int\limits_{R^d} (g (s+t, y) - g (s, x)) P (s, x, t+s, dy)$

by means of a Taylor expansion of $g (s+t, y)$ about (s, x) under the assumption that g is defined and bounded on $[t_0, T] \times R^d$ and is, on that set, twice continuously differentiable with respect to the x_i and once continuously differentiable with respect to s. When we use conditions b) and c) of definition (2.5.1), we obtain for the right-hand member of (2.6.2) the operator $\partial / \partial s + \mathfrak{D}$. Under certain conditions on f and B (which we shall specify in section 9.4) we have, for all functions in D_A,

$$A = \frac{\partial}{\partial s} + \mathfrak{D}$$

or, for time-independent functions and the homogeneous case,

$$A = \mathfrak{D}.$$

Therefore, the diffusion process is uniquely determined in this case by f and B. Furthermore, we see that the first derivatives in $\mathfrak{D}$ arise as a result of systematic drift and the second derivatives as a result of the local irregular "chaotic" fluctuational motions.

In the next chapter, we shall take a purely probabilistic route to the construction of a diffusion process for a given operator. A purely analytical approach yields

(2.6.3)   **Theorem.** Let X_t, for $t_0 \leqq t \leqq T$, denote a d-dimensional diffusion process with continuous coefficients $f (s, x)$ and $B (s, x)$. The limit relations in definition (2.5.1) hold uniformly in $s \in [t_0, T)$. Let $g (x)$ denote a continuous bounded scalar function such that

$$u (s, x) = E_{s, x} g (X_t) = \int\limits_{R^d} g (y) P (s, x, t, dy)$$

for $s < t$, where t is fixed, and $x \in R^d$ is continuous and bounded, as are its derivatives $\partial u / \partial x_i$ and $\partial^2 u / \partial x_i \, \partial x_j$ for $1 \leqq i, j \leqq d$. Then, $u (s, x)$ is differentiable with respect to s and satisfies **Kolmogorov's backward equation**

(2.6.4) $\dfrac{\partial u}{\partial s} + \mathfrak{D} u = 0,$

where $\mathfrak{D}$ is the operator (2.6.1), with the end condition

$$\lim\limits_{s \uparrow t} u (s, x) = g (x).$$

A proof of this theorem can again be obtained by means of a Taylor expansion of u. For details, see Gikhman and Skorokhod [5], p. 373. The name "backward equation" stems from the fact that the differentiation is with respect to the backward time arguments s and x in contrast with the forward equation (see Theorem (2.6.9)), in which the transition density $p(s, x, t, y)$ is differentiated with respect to t and y.

(2.6.5) **Remark.** Theoretically, the backward equation (2.6.4) enables us to determine the transition probability $P(s, x, t, \cdot)$. This transition probability is uniquely defined if we know all the integrals

$$u(s, x) = \int_{R^d} g(y)\, P(s, x, t, dy),$$

where g ranges over a set of functions that is dense in the space $C(R^d)$ of continuous bounded functions. If the solution of (2.6.4) is unique for these functions g, we can, for known f and B, calculate $u(s, x)$ from it and then calculate $P(s, x, t, \cdot)$.

(2.6.6) **Theorem.** Suppose that the assumptions of Theorem (2.6.3) regarding X_t hold. If $P(s, x, t, \cdot)$ has a density $p(s, x, t, y)$ that is continuous with respect to s and if the derivatives $\partial p/\partial x_i$ and $\partial^2 p/\partial x_i\, \partial x_j$ exist and are continuous with respect to s, then p is a so-called **fundamental solution** of the backward equation

$$\frac{\partial p}{\partial s} + \mathfrak{D}\, p = 0;$$

that is, it satisfies the end condition

$$\lim_{s \uparrow t} p(s, x, t, y) = \delta(x - y),$$

where δ is Dirac's delta function.

(2.6.7) **Example.** The transition density of the Wiener process

$$p(s, x, t, y) = (2\pi(t - s))^{-d/2}\, e^{-|y - x|^2/2(t - s)}$$

is, for fixed t and y, a fundamental solution of the backward equation

$$\frac{\partial p}{\partial s} + \frac{1}{2} \sum_{i=1}^{d} \frac{\partial^2 p}{\partial x_i^2} = 0.$$

(2.6.8) **Remark.** If X_t is a *homogeneous* process, then the coefficients $f(s, x) \equiv f(x)$ and $B(s, x) \equiv B(x)$ (and hence the operator $\mathfrak{D}$) are independent of s. Since $P(s, x, t, B) = P(t - s, x, B)$, the sign of $\partial p/\partial s$ changes in the backward equation, for example, for the density $p(s, x, y)$; that is,

$$-\frac{\partial p}{\partial s} + \mathfrak{D}\, p = 0.$$

(2.6.9) **Theorem.** Let X_t, for $t_0 \leq t \leq T$, denote a d-dimensional diffusion process for which the limit relationships in definition (2.5.1) hold uniformly in s

and x and which possesses a transition density $p(s, x, t, y)$. If the derivatives $\partial p/\partial t$, $\partial (f_i(t, y) p)/\partial y_i$ and $\partial^2 (b_{ij}(t, y) p)/\partial y_i \partial y_j$ exist and are continuous functions, then, for fixed s and x such that $s \leqq t$, this transition density $p(s, x, t, y)$ is a fundamental solution of **Kolmogorov's forward** or the **Fokker-Planck equation**

$$(2.6.10) \quad \frac{\partial p}{\partial t} + \sum_{i=1}^{d} \frac{\partial}{\partial y_i} (f_i(t, y) p) - \frac{1}{2} \sum_{i=1}^{d} \sum_{j=1}^{d} \frac{\partial^2}{\partial y_i \partial y_j} (b_{ij}(t, y) p) = 0.$$

For proof, we again refer to Gikhman and Skorokhod [5], p. 375.

If we define the distribution of X_{t_0} in terms of the initial probability P_{t_0}, we obtain from $p(s, x, t, y)$ the probability density $p(t, y)$ of X_t itself:

$$p(t, y) = \int_{R^d} p(t_0, x, t, y) P_{t_0}(dx).$$

If we apply the integration with respect to $P_{t_0}(dx)$ to (2.6.10), we see that $p(t, y)$ also satisfies the forward equation.

(2.6.11) **Example.** For the Wiener process, the forward equation for the homogeneous transition density

$$p(t, x, y) = (2 \pi t)^{-d/2} e^{-|y-x|^2/2t}$$

becomes

$$\frac{\partial p}{\partial t} = \frac{1}{2} \sum_{i=1}^{d} \frac{\partial^2 p}{\partial y_i^2},$$

which in this case is identical to the backward equation with x replaced by y.

Chapter 3
Wiener Process and White Noise

3.1 Wiener Process

Let us now summarize the most important properties of the d-dimensional Wiener process W_t defined in section 2.3. This process, a mathematical model of Brownian motion of a free particle with friction neglected, is a spacewise and timewise homogeneous diffusion process with drift vector $f \equiv 0$ and diffusion matrix $B \equiv I$. A clear understanding of this process is especially important since it proves to be the fundamental building block for all (smooth) diffusion processes. We have already seen this in the heuristic derivation of the stochastic differential equation of section 2.5.

Since W_t is a Markov process, all the distributions of W_t are defined, in accordance with (2.2.6), by the initial condition

$$W_0 = 0$$

and the stationary transition density

$$p(t, x, y) = n(t, x, y) = (2\pi t)^{-d/2} \exp(-|y - x|^2 / 2t).$$

From this we get the density $p(t, y)$ of W_t itself:

$$p(t, y) = n(t, y) = n(t, 0, y) = (2\pi t)^{-d/2} \exp(-|y|^2 / 2t).$$

This is the density of the d-dimensional normal distribution $\mathfrak{N}(0, t\,I)$. The n-dimensional distribution of W_t, $P[W_{t_1} \in B_1, \ldots, W_{t_n} \in B_n]$, where $0 < t_1 < \ldots < t_n$, then has, according to formula (2.2.6), the density

$$(3.1.1) \qquad n(t_1, 0, x_1)\, n(t_2 - t_1, x_1, x_2) \ldots n(t_n - t_{n-1}, x_{n-1}, x_n).$$

Since $E\,|W_t - W_s|^4 = (d^2 + 2d)(t - s)^2$, it follows immediately from Kolmogorov's criterion (see section 1.8) that for these finite-dimensional distributions there exists a stochastic process with continuous sample functions. Since

$$n(t, x, y) = \prod_{i=1}^{d} (2\pi t)^{-1/2} \exp(-|y_i - x_i|^2 / 2t),$$

the d components of W_t are themselves independent one-dimensional Wiener processes.

If $\mathfrak{W}_t = \mathfrak{A}(W_s, s \leq t)$, then, for $s \leq t$,

$$E(W_t | \mathfrak{W}_s) = E(W_t | W_s) = \int_{R^d} x \, n(t-s, W_s, x) \, dx = W_s;$$

that is, W_t is a d-dimensional martingale.

Of basic importance is the following property: W_t *has independent increments;* that is, for $0 < t_1 < \dots < t_n$, the random variables

$$W_{t_1}, W_{t_2} - W_{t_1}, \dots, W_{t_n} - W_{t_{n-1}}$$

are independent. Here, $W_t - W_s$ (for $s < t$) has the distribution $\mathfrak{N}(0, (t-s)I)$, which depends only on $t-s$; that is, the increments are **stationary**. The last two assertions follow immediately from (3.1.1) when we remember that, in our case, the value of $n(s, x, t, y) = n(t-s, x, y)$ depends only on $y - x$ and $t - s$.

In fact, a Wiener process can be *defined* as a process with independent and stationary $\mathfrak{N}(0, (t-s)I)$-distributed increments $W_t - W_s$, with initial value $W_0 = 0$, and with almost certainly continuous sample functions.

The fact that W_t is a process with independent and stationary increments makes it possible to apply the limit theorems for sums of independent identically distributed random variables. This provides valuable information regarding the order of magnitude of the sample functions of W_t. The **strong law of large numbers** states that

$$\lim_{t \to \infty} \frac{W_t}{t} = 0 \quad \text{with probability 1.}$$

The true order of magnitude of the sample functions follows from the **law of the iterated logarithm**: for $d = 1$ (that is, for each individual component of a d-dimensional Wiener process),

$$\limsup_{t \to \infty} \frac{W_t}{\sqrt{2t \log \log t}} = 1$$

and

$$\liminf_{t \to \infty} \frac{W_t}{\sqrt{2t \log \log t}} = -1,$$

both with probability 1. This means that, for every $\varepsilon > 0$ and for almost every sample function $W_t(\omega)$, there exists an instant $t_0(\omega)$ subsequent to which we always have

$$-(1+\varepsilon)\sqrt{2t \log \log t} \leq W_t(\omega) \leq (1+\varepsilon)\sqrt{2t \log \log t}.$$

On the other hand, the bounds $(1-\varepsilon)\sqrt{2t \log \log t}$ and $-(1-\varepsilon)\sqrt{2t \log \log t}$

(for $0 < \varepsilon < 1$) are exceeded in every t-neighborhood of ∞ for almost every sample function.

For a d-dimensional Wiener process, we have in general

$$(3.1.2) \qquad \lim_{t \to \infty} \sup \frac{|W_t|}{\sqrt{2\,t \log \log t}} = 1,$$

which is somewhat surprising in that it means that the individual (independent!) components of W_t are *not simultaneously* of the order $\sqrt{2\,t \log \log t}$. This is true because, otherwise, $\sqrt{d}$ would appear in the right-hand member of (3.1.2). For a generalization, see Theorem (7.2.5).

Equation (3.1.2) becomes plausible by virtue of the invariance of W_t under rotation. We have

(3.1.3) **Lemma.** a) W_t is a Gaussian stochastic process with expectation $E\,W_t$ = 0 and covariance matrix

$$E\,W_t\,W_s' = \min\,(t, s)\,I.$$

b) W_t is invariant under rotations in R^d; that is, if W_t is a Wiener process, so is $V_t = U\,W_t$, where U is an orthogonal matrix.

c) If W_t is a Wiener process, the processes $-W_t$, $c\,W_{t/c^2}$ (where $c \neq 0$), $t\,W_{1/t}$, and $W_{t+s} - W_s$ (where s is fixed and $t \geq 0$) are also Wiener processes.

Having Lemma (3.1.3c) at our disposal, we are in the position to study also the local behavior of W_t.

The application of the law of the iterated logarithm to $t\,W_{1/t}$ yields, for $d = 1$ (and hence for every component of a d-dimensional Wiener process),

$$\lim_{t \downarrow 0} \sup \frac{W_t}{\sqrt{2\,t \log \log 1/t}} = 1$$

and

$$\lim_{t \downarrow 0} \inf \frac{W_t}{\sqrt{2\,t \log \log 1/t}} = -1,$$

for d-dimensional W_t

$$\lim_{t \downarrow 0} \sup \frac{|W_t|}{\sqrt{2\,t \log \log 1/t}} = 1,$$

for almost all sample functions. One consequence of this is that every component of sample functions W_t has, with probability 1, in every interval of the form $[0, \varepsilon)$ with $\varepsilon > 0$ infinitely many zeros, which cluster about the point $t = 0$. This behavior is exhibited at every point $s \geq 0$ because, by Lemma

(3.1.3), part c), when W_t is a Wiener process, $W_{t+s} - W_s$ (for fixed s and non-negative t) is also a Wiener process (independent in fact of W_t for $t \leqq s$).

Almost all sample functions of a Wiener process are continuous though, in accordance with a theorem of N. Wiener, *nowhere differentiable* functions. Proof of this assertion and most of the previously made assertions regarding W_t can be found, for example, in McKean [45]. For fixed t, the nondifferentiability can be made clear as follows: The distribution of the difference quotient $(W_{t+h} - W_t)/h$ is $\mathfrak{N}(0, (1/|h|)I)$. As $h \rightarrow 0$, this normal distribution diverges, so that, for every bounded measurable set B,

$$P[(W_{t+h} - W_t)/h \in B] \rightarrow 0 \quad (h \rightarrow 0).$$

Therefore, the difference quotient cannot converge with positive probability to a finite random variable.

We can get more precise information from the law of the iterated logarithm. For $d = 1$ (hence for every individual component of a many-dimensional process), we obtain for almost every sample function and arbitrary ε in the interval $0 < \varepsilon < 1$, as $h \downarrow 0$,

$$\frac{W_{t+h} - W_t}{h} \geqq (1 - \varepsilon) \sqrt{\frac{2 \log \log 1/h}{h}} \quad \text{infinitely often}$$

and, simultaneously,

$$\frac{W_{t+h} - W_t}{h} \leqq (-1 + \varepsilon) \sqrt{\frac{2 \log \log 1/h}{h}} \quad \text{infinitely often.}$$

Since the right-hand members approach $+\infty$ and $-\infty$ respectively as $h \downarrow 0$, the ratio $(W_{t+h} - W_t)/h$ has with probability 1, for every fixed t, the extended real line $[-\infty, +\infty]$ as its set of cluster points.

For the case of Brownian motion described by a Wiener process, the nondifferentiability means that the particle under observation does not possess a velocity at any instant. This disadvantage is offset in a model treated in section 8.3, namely, the Ornstein-Uhlenbeck process.

The local law of the iterated logarithm reveals therefore enormous local fluctuations in W_t. The crucial property as regards the difficulties in the definition of a Steiltjes integral with respect to W_t is that each portion of almost every sample function of W_t is of unbounded variation in a finite interval of time; that is, its length is infinite. This is a consequence of the following more precise result:

(3.1.4) **Lemma.** Let W_t denote a d-dimensional Wiener process and let $s = t_0^{(n)} < t_1^{(n)} < \ldots < t_n^{(n)} = t$ denote a sequence of decompositions of the interval $[s, t]$ such that $\delta_n = \max(t_k^{(n)} - t_{k-1}^{(n)})$. Then (we write t_k for $t_k^{(n)}$ for short),

(3.1.5) $\text{qm-}\lim\limits_{\delta_n \to 0} \sum\limits_{k=1}^{n} (W_{t_k} - W_{t_{k-1}})(W_{t_k} - W_{t_{k-1}})' = (t-s) I,$

where elementwise convergence of the matrices is meant. In particular,

(3.1.6) $\text{qm-}\lim\limits_{\delta_n \to 0} \sum\limits_{k=1}^{n} |W_{t_k} - W_{t_{k-1}}|^2 = d(t-s).$

If δ_n approaches 0 so fast that $\sum \delta_n < \infty$, then convergence occurs in (3.1.5) and (3.1.6) also with probability 1.

Proof. Let W_t^i, for $i = 1, \ldots, d$, denote the ith component of W_t. If

$$S_n^{ij} = \sum_{k=1}^{n} (W_{t_k}^i - W_{t_{k-1}}^i)(W_{t_k}^j - W_{t_{k-1}}^j),$$

then,

$$E(S_n^{ij}) = (t-s)\delta_{ij}$$

and

$$V(S_n^{ij}) = \sum_{k=1}^{n} (E(W_{t_k}^i - W_{t_{k-1}}^i)^2 (W_{t_k}^j - W_{t_{k-1}}^j)^2 - \delta_{ij}(t_k - t_{k-1})^2)$$

$$= (1 + \delta_{ij}) \sum_{k=1}^{n} (t_k - t_{k-1})^2$$

$$\leq 2(t-s)\delta_n \to 0 \quad (\delta_n \to 0),$$

which proves (3.1.5). If we apply the trace operator to both sides of (3.1.5), we obtain (3.1.6). If $\sum \delta_n < \infty$, then $\sum\limits_{n} V(S_n^{ij}) < \infty$, which, by virtue of section 1.4, is sufficient for almost certain convergence in (3.1.5). ■

Let us look at the decomposition with intermediate points $t_k^{(n)} = s + (t-s) k/2^n$, for $k = 0, 1, \ldots, 2^n$ and $n = 1, 2, \ldots$. Since $\delta_n = (t-s) 2^{-n}$ and $\sum \delta_n < \infty$, the left-hand member of the inequality

$$\sum_{k=1}^{2^n} |W_{t_k} - W_{t_{k-1}}|^2 \leq \max_{k=1,\ldots,2^n} |W_{t_k} - W_{t_{k-1}}| \sum_{k=1}^{2^n} |W_{t_k} - W_{t_{k-1}}|$$

converges, for almost every sample function, to the finite random variable $d(t-s)$ as $n \to \infty$. The almost certain continuity of sample functions implies that

$$\max_{k=1,\ldots,2^n} |W_{t_k} - W_{t_{k-1}}| \to 0 \quad (n \to \infty),$$

and this implies that

Fig. 4:
Sample function of
the Wiener process.

$$\sum_{k=1}^{2^n} |W_{t_k} - W_{t_{k-1}}| \to \infty \quad (n \to \infty)$$

with probability 1; that is, almost all sample functions of W_t are of unbounded variation in every finite interval.

(3.1.7) Remark. Equations (3.1.5) and (3.1.6) serve as motivation for the symbolic notation (frequently used, especially in the case $d = 1$)

$$(dW_t)(dW_t') = I \, dt$$

and, for $d = 1$,

$$(dW_t)^2 = dt.$$

3.2 White Noise

We confine ourselves for the moment to the one-dimensional case. So-called (Gaussian) **white noise** is generally understood in engineering literature as a stationary Gaussian process ξ_t, for $-\infty < t < \infty$, with mean $E \, \xi_t = 0$ and a constant spectral density $f(\lambda)$ on the entire real axis. If $E \, \xi_s \, \xi_{t+s} = C(t)$ is the covariance function of ξ_t, then

$$(3.2.1) \qquad f(\lambda) = \frac{1}{2\pi} \int_{-\infty}^{\infty} e^{-i\lambda t} \, C(t) \, dt = \frac{c}{2\pi} \qquad \text{for all } \lambda \in R^1 \, ,$$

where c is a positive constant, which we can, without loss of generality, take equal to 1.

Therefore, such a process has a spectrum on which all frequencies participate with the same intensity, hence a "white" spectrum (in analogy with "white light" in optics, which contains all frequencies of visible light uniformly). However, such a process does not exist in the traditional sense because (3.2.1) is compatible only with the choice

$$C(t) = \delta(t),$$

where δ is Dirac's delta function. In particular, we would have

$$C(0) = E\,\xi_t^2 = \int\limits_{-\infty}^{\infty} f(\lambda)\,d\lambda = \infty.$$

Since $C(t) = 0$ for $t \neq 0$, the values of ξ_s and ξ_{s+t} would be uncorrelated for arbitrarily small values of t (and independent, in fact, since the process is Gaussian), a fact that explains the name "purely random process". Obviously, the sample functions of a process with independent values at all instants must be extremely irregular.

In fact, white noise was first correctly described in connection with the theory of generalized functions (distributions), as is done, for example, in Gel'fand and Vilenkin [22], Chapter III. Let us discuss this matter briefly.

We start with the fact that, in every actual measurement of the values of a function $f(t)$, the inertia of the measuring instrument allows us to get only an average value

(3.2.2) $$\Phi_f(\varphi) = \int\limits_{-\infty}^{\infty} \varphi(t) f(t)\,dt,$$

where $\varphi(t)$ is a function characterizing the measuring instrument. The function Φ_f depends linearly and continuously on φ. It is the generalized function corresponding to f.

As a consequence of the smoothing effect of the measuring instrument, we obtain a value for the integral (3.2.2) even when the values of f do not actually exist at individual points. This leads to the following general definition:

Let K denote the space of all infinitely differentiable functions $\varphi(t)$, for $t \in R^1$, that vanish identically outside a finite interval (which in general depends on $\varphi(t)$). A sequence $\varphi_1(t)$, $\varphi_2(t)$, ... of such functions is said to converge to $\varphi(t) \equiv 0$ if all these functions vanish outside a single bounded region and if all of them and all of their derivatives (in the usual sense) converge uniformly to 0. Every continuous linear functional Φ defined on the space K is called a **generalized function** (or a **distribution*). The generalized function defined by

$$\Phi(\varphi) = \varphi(t_0) \text{ for all} \varphi \in K, \ t_0 \in R^1 \text{ fixed,}$$

is called the Dirac delta function and is denoted by $\delta(t - t_0)$. In contrast with classical functions, generalized functions always have derivatives of every order, which again are generalized functions. By the derivative $\dot{\Phi}$ of Φ, we mean the generalized function defined by

$$\dot{\Phi}(\varphi) = -\Phi(\dot{\varphi}).$$

---

*Henceforth, we shall not use this term in this sense, so that whenever "distributions" are mentioned, they have the probability-theory meaning.

A **generalized stochastic process** is now simply a random generalized function in the following sense: to every $\varphi \in K$ is assigned a random variable $\Phi(\varphi)$ (in other words, $\Phi(\varphi)$ is an ordinary stochastic process with parameter set K) such that the following two conditions hold:

1.   The functional Φ is *linear* on K with probability 1; that is, for arbitrary φ and ψ in K and arbitrary numbers α and β, we have

$$\Phi(\alpha\,\varphi + \beta\,\psi) = \alpha\,\Phi(\varphi) + \beta\,\Phi(\psi)$$

with probability 1;

2. $\Phi(\varphi)$ is *continuous* in the following sense: The convergence of the functions φ_{kj} to φ_k in the space K, for $k = 1, 2, \ldots, n$, as $j \to \infty$ implies convergence of the distribution of the vector $(\Phi(\varphi_{1j}), \ldots, \Phi(\varphi_{nj}))$ to the distribution of $(\Phi(\varphi_1), \ldots, \Phi(\varphi_n))$ in the sense of distribution convergence defined in section 1.4.

For example, some generalized stochastic process corresponds to every ordinary stochastic process with continuous sample functions via formula (3.2.2).

A generalized stochastic process is said to be **Gaussian** if, for arbitrary linearly independent functions $\varphi_1, \ldots, \varphi_n \in K$, the random variable $(\Phi(\varphi_1), \ldots, \Phi(\varphi_n))$ is normally distributed. Just as in the classical case, a generalized Gaussian process is uniquely defined by the continuous linear mean-value functional

$$E\,\Phi(\varphi) = m(\varphi)$$

and the continuous bilinear positive-definite covariance functional

$$E\,(\Phi(\varphi) - m(\varphi))(\Phi(\psi) - m(\psi)) = C(\varphi, \psi).$$

One of the important advantages of a generalized stochastic process is the fact that its derivative always exists and is itself a generalized stochastic process. In fact the derivative $\dot{\Phi}$ of Φ is the process defined by setting

$$\dot{\Phi}(\varphi) = -\Phi(\dot{\varphi}).$$

The derivative of a Gaussian process with mean $m(\varphi)$ and covariance $C(\varphi, \psi)$ is again a Gaussian process and it has mean value $\dot{m}(\varphi) = -m(\dot{\varphi})$ and covariance $C(\dot{\varphi}, \dot{\psi})$.

As an example, let us look at a Wiener process and its derivative. From the representation

$$\Phi(\varphi) = \int\limits_{-\infty}^{\infty} \varphi(t)\, W_t\, dt$$

(we set $W_t \equiv 0$ for $t < 0$), we conclude immediately that, with W_t regarded as a generalized Gaussian stochastic process, we have

$$m(\varphi) \equiv 0$$

and

$$C(\varphi, \psi) = \int\limits_0^\infty \int\limits_0^\infty \min(t, s) \, \varphi(t) \, \psi(s) \, dt \, ds.$$

After some elementary manipulations and integration by parts, we get

$$C(\varphi, \psi) = \int\limits_0^\infty (\widehat{\varphi}(t) - \widehat{\varphi}(\infty))(\widehat{\psi}(t) - \widehat{\psi}(\infty)) \, dt,$$

where

$$\widehat{\varphi}(t) = \int\limits_0^t \varphi(s) \, ds, \quad \widehat{\psi}(t) = \int\limits_0^t \psi(s) \, ds.$$

Let us now calculate the derivative of the Wiener process. This is a generalized Gaussian stochastic process with mean value $\dot{m}(\varphi) \equiv 0$ and covariance

$$\dot{C}(\varphi, \psi) = C(\dot{\varphi}, \dot{\psi})$$

$$= \int\limits_0^\infty \varphi(t) \, \psi(t) \, dt.$$

This formula can be put in the form

$$\dot{C}(\varphi, \psi) = \int\limits_0^\infty \int\limits_0^\infty \delta(t - s) \, \varphi(t) \, \psi(t) \, dt.$$

Therefore, the covariance function of the derivative of the Wiener process is the generalized function

$$\dot{C}(s, t) = \delta(t - s).$$

But this is the covariance function of white noise! Thus, *white noise ξ_t is the derivative of the Wiener process W_t* when we consider both processes as generalized stochastic processes. This justifies the notation

(3.2.3a) $\quad \xi_t = \dot{W}_t$

frequently used in engineering literature. Of course, we also have conversely

(3.2.3b) $\quad W_t = \int\limits_0^t \xi_s \, ds$

in the sense of coincidence of the covariance functionals.

By virtue of what we have said, we can now give the *definition*: a Gaussian white noise ξ_t, for $t \in R^1$, is a generalized Gaussian stochastic process Φ_ξ with mean value 0 and covariance functional

(3.2.4) $\qquad C_\xi(\varphi, \psi) = \int\limits_{-\infty}^{\infty} \varphi(t)\, \psi(t)\, \mathrm{d}t.$

From (3.2.4), we conclude that

$$C_\xi(\varphi(t), \psi(t)) = C_\xi(\varphi(t+h), \psi(t+h)), \quad h \in R^1,$$

a consequence of which is that, for arbitrary functions $\varphi_1, \ldots, \varphi_n \in K$, the random variable $(\Phi_\xi(\varphi_1(t+h)), \ldots, \Phi_\xi(\varphi_n(t+h))$ has the same distribution for all h; that is, white noise is a *stationary* generalized process. One can show that, up to a factor, the spectral measure of this generalized process is Lebesgue measure and hence the process has a constant spectral density on the entire real axis.

We also conclude from (3.2.4) that

$$C_\xi(\varphi, \psi) = 0 \quad \text{if} \quad \varphi(t)\cdot\psi(t) \equiv 0;$$

that is, the random variables $\Phi_\xi(\varphi)$ and $\Phi_\xi(\psi)$ are independent in this case. We say that a generalized stochastic process with this property *has at every point independent values*. The class of stationary generalized processes with independent values at every point is well known (see Gel'fand and Vilenkin [22]). Roughly speaking, we might say that they are obtained by differentiation of processes with stationary and independent increments. All these processes can serve as models of "noise", that is, of stationary and rapidly fluctuating phenomena. The "noise" is "white" (that is, the spectral density is constant) if the covariance functional has the form (3.2.4). Although in the present book we shall consider Gaussian white noise exclusively, there exist yet other important (non-Gaussian) white noise processes, for example, so-called **Poisson white noise**, which represents the derivative of a Poisson process (after subtraction of the mean value).

Thus, although a stationary Gaussian process ξ_t with everywhere constant spectral density in the traditional sense does not exist, such a concept nonetheless proves to be a very useful mathematical idealization. Furthermore, we conclude from equations (3.2.3a) and (3.2.3b) that ξ_t is, so to speak, only a derivation of a classical stochastic process and that, therefore, only the smoothing effect of a single integration is needed to return from ξ_t to an ordinary process, namely, W_t. This last is also the reason for converting differential equations containing white noise into integral equations.

For treatment of integrals of the form

$$\int\limits_{-\infty}^{\infty} f(t)\, \xi_t\, \mathrm{d}t,$$

where ξ_t is now an arbitrary generalized stochastic process with independent values at every point, we refer to the paper by Dawson [34].

Because of the independence of the values at every point, white noise is appropriate for describing rapidly fluctuating random phenomena, for which the correlation of the state at the instant t with the state at the time s when $|t-s|$ is increasing becomes small very rapidly. For example, this is the case with the force acting on the particle observed in the case of Brownian motion or for the variation in current in an electrical circuit due to thermal noise.

White noise ξ_t can be approximated by an ordinary stationary Gaussian process X_t, for example, one with covariance

$$C(t) = a\, e^{-b|t|} \quad (a>0,\, b>0).$$

Such a process has spectral density

$$f(\lambda) = \frac{1}{\pi} \frac{a\,b}{b^2 + \lambda^2}.$$

If we now let a and b approach ∞ in such a way that $a/b \to 1/2$, we get

$$f(\lambda) \to \frac{1}{2\pi} \quad \text{for all} \quad \lambda \in R^1$$

and

$$C(t) \to \begin{cases} 0, & t \neq 0, \\ \infty, & t = 0, \end{cases}$$

but

$$\int_{-\infty}^{\infty} C(t)\, dt \to 1,$$

so that

$$C(t) \to \delta(t);$$

that is, X_t converges in a certain sense to ξ_t.

Let us now look at the indefinite integral

$$Y_t = \int_0^t X_s\, ds.$$

This is again a Gaussian process with $E\, Y_t = 0$ and covariance

$$E\, Y_t Y_s = \int_0^t \int_0^s a\, e^{-b|u-v|}\, du\, dv.$$

Taking the limit as above, we get

$$E\, Y_t Y_s \to \min(t, s),$$

that is, the covariance of the one-dimensional Wiener process W_t. This is a further heuristic justification of formulas (3.2.3).

Now, we can define the d-**dimensional (Gaussian) white noise** as the derivative (in the generalized function sense) of the d-dimensional Wiener process. It is a stationary Gaussian generalized process with independent values at every point, with expectation vector 0, and with covariance matrix $\delta(t) I$. In other words, white noise in R^d is simply a combination of d independent one-dimensional white noise processes. The spectral density (now a matrix!) of such a process is $I/2\pi$.

A d-dimensional Gaussian noise process with expectation 0 and with covariance matrix

$$E\,\eta_t\,\eta_s' = Q(t)\,\delta(t-s)$$

is treated by various authors (for example, Bucy and Joseph [61] and Jazwinski [66]). Such a process is no longer in general stationary but, as a "delta-correlated" process, it has independent values at every point. We shall see later (see remarks (5.2.4) and (5.4.8)) that we can confine ourselves to the standard case $Q(t) = I$ without loss of generality. We obtain η_t from the standard noise process ξ_t by

$$\eta_t = G(t)\,\xi_t,$$

where $G(t)$ is any $(d \times d)$-matrix-valued function such that $G(t)\,G(t)' = Q(t)$.

Chapter 4
Stochastic Integrals

4.1 Introduction

The analysis of stochastic dynamic systems often leads to differential equations of the form

(4.1.1) $\qquad \dot{X}_t = f(t, X_t) + G(t, X_t)\, \xi_t \, ,$

where we can assume that ξ_t is white noise. Here, X_t and f are R^d-valued functions, $G(t, x) = (G_{ij}(t, x))$ is a $d \times m$ matrix and ξ_t is m-dimensional white noise.

We saw in section 3.2 that, although ξ_t is not a usual stochastic process, nonetheless the indefinite integral of ξ_t can be identified with the m-dimensional Wiener process W_t:

$$W_t = \int\limits_0^t \xi_s \, ds$$

or, in shorter symbolic notation,

$$dW_t = \xi_t \, dt.$$

The solution of a deterministic initial-value problem

$$\dot{x}_t = f(t, x_t), \quad x_{t_0} = c,$$

for a continuous function $f(t, x)$ is, as we know, equivalent to the solution of the integral equation

$$x_t = c + \int\limits_{t_0}^t f(s, x_s) \, ds,$$

for which it is possible to find a solution curve by means of the classical iteration procedure.

In the same way, we transform equation (4.1.1) into an integral equation

(4.1.2) $\qquad X_t = c + \int\limits_{t_0}^t f(s, X_s) \, ds + \int\limits_{t_0}^t G(s, X_s) \, \xi_s \, ds.$

Here, c is an arbitrary random variable, which can also degenerate into a constant independent of chance. As a rule, the first integral in the right-hand member of equation (4.1.2) can be understood as the familiar Riemann integral. The second integral is more of a problem. Because of the smoothing effect of the integration, we still hope to be able to interpret integrals of this form for many functions $G(t, x)$ as ordinary random variables, which would spare us necessity of using generalized stochastic processes. We now formally eliminate the white noise in (4.1.2) by means of the relationship $dW_t = \xi_t \, dt$, writing

$$(4.1.3) \qquad \int_{t_0}^{t} G(s, X_s) \, \xi_s \, ds = \int_{t_0}^{t} G(s, X_s) \, dW_s,$$

so that (4.1.2) takes the form

$$(4.1.4) \qquad X_t = c + \int_{t_0}^{t} f(s, X_s) \, ds + \int_{t_0}^{t} G(s, X_s) \, dW_s \, .$$

Equation (4.1.4) is also written more briefly in the following differential form:

$$(4.1.5) \qquad dX_t = f(t, X_t) \, dt + G(t, X_t) \, dW_t.$$

Since, in accordance with section 3.1, almost all sample functions of W_t are of unbounded variation, we cannot in general interpret the integral in the right-hand member of (4.1.4) as an ordinary Riemann-Stieltjes integral. We shall consider this with the example given in the following section.

For fixed (i.e., independent of ω) continuously differentiable functions g, we could use the formula for integration by parts to give the following definition:

$$(4.1.6) \qquad \int_{t_0}^{t} g(s) \, dW_s = g(t) \, W_t - g(t_0) \, W_{t_0} - \int_{t_0}^{t} \dot{g}(s) \, W_s \, ds.$$

The last integral is an ordinary Riemann integral, evaluated for the individual sample functions of W_t .

In many cases of importance in practice, however, the function $G(t, x)$ in the integral equation (4.1.2) is not independent of x. For this general case, K. Itô [42] has given a definition of the integral (4.1.3) that, as we shall see, includes the definition (4.1.6) as a special case.

4.2 An Example

Our task is now to define the integral

$$X_t = X_t(\omega) = \int_{t_0}^{t} G(s) \, dW_s = \int_{t_0}^{t} G(s, \omega) \, dW_s(\omega) \, (0 \leqq t_0 < t, \text{ fixed})$$

for as broad a class of ($d \times m$ matrix)-valued random functions G as possible. Here, W_t is an m-dimensional Wiener process.

If $G\left(\cdot,\omega\right)$ is, for almost all ω, fairly smooth, it cannot, so to speak, penetrate the local irregularities of the sample functions $W.\left(\omega\right)$ and exhaust them.

For example, for $G \equiv 1$ and $d = m = 1$, we set

$$\int_{t_0}^{t} 1 \, dW_s = W_t - W_{t_0}$$

since every approximation of the integral by means of Riemann-Stieltjes sums of the form

$$S_n = \sum_{i=1}^{n} G\left(\tau_i\right)\left(W_{t_i} - W_{t_{i-1}}\right), \quad t_0 \leq t_1 \leq \ldots \leq t_n = t, \quad t_{i-1} \leq \tau_i \leq t_i,$$

leads to this value.

The situation is different when G is about as irregular as W_t itself. To this end, let us look, as an example, at the case $G\left(t\right) = W_t$, $d = m = 1$, so that the integral is

$$X_t = \int_{t_0}^{t} W_s \, dW_s \,.$$

Formal application of the classical rules for integration by parts yields

(4.2.1) $$\int_{t_0}^{t} W_s \, dW_s = \left(W_t^2 - W_{t_0}^2\right)/2 \,.$$

However, this calculation assumes the existence of the integral as an ordinary Riemann-Stieltjes integral; that is, it assumes convergence of the sums

$$S_n = \sum_{i=1}^{n} W_{\tau_i}\left(W_{t_i} - W_{t_{i-1}}\right)$$

with ever finer partitioning and *arbitrary* choice of the intermediate points τ_i. Let us now show that the limit of $\{S_n\}$ depends on the choice of the intermediate points.

To do this, let us write the sum S_n in the form

$$S_n = W_t^2/2 - W_{t_0}^2/2 - \frac{1}{2} \sum_{i=1}^{n} \left(W_{t_i} - W_{t_{i-1}}\right)^2$$

$$+ \sum_{i=1}^{n} \left(W_{\tau_i} - W_{t_{i-1}}\right)^2 + \sum_{i=1}^{n} \left(W_{t_i} - W_{\tau_i}\right)\left(W_{\tau_i} - W_{t_{i-1}}\right).$$

In accordance with Lemma (3.1.1) with $\delta_n = \max_i (t_i - t_{i-1})$,

$$\underset{\delta_n \to 0}{\text{qm-lim}} \sum_{i=1}^{n} (W_{t_i} - W_{t_{i-1}})^2 = t - t_0.$$

By calculating the two first moments, we can easily show that

$$\underset{\delta_n \to 0}{\text{qm-lim}} \sum_{i=1}^{n} (W_{t_i} - W_{\tau_i}) (W_{\tau_i} - W_{t_{i-1}}) = 0.$$

For the remaining sum, we have

(4.2.2) $$E \sum_{i=1}^{n} (W_{\tau_i} - W_{t_{i-1}})^2 = \sum_{i=1}^{n} (\tau_i - t_{i-1})$$

and

$$V\left(\sum_{i=1}^{n} (W_{\tau_i} - W_{t_{i-1}})^2 \right) = 2 \sum_{i=1}^{n} (\tau_i - t_{i-1})^2$$

$$\leqq 2 (t - t_0) \delta_n \longrightarrow 0 \quad (\delta_n \to 0).$$

Therefore, the convergence of $\{S_n\}$ depends on the behavior of the sums (4.2.2), which can assume any value in the interval $[0, t - t_0]$ with appropriate choice of the τ_i. More precisely,

(4.2.3) $$\underset{\delta_n \to 0}{\text{qm-lim}} \left(S_n - \sum_{i=1}^{n} (\tau_i - t_{i-1}) \right) = (W_t^2 - W_{t_0}^2)/2 - (t - t_0)/2.$$

Therefore, in order to obtain a unique definition of the integral, it is necessary to *define* specific intermediate points τ_i. Of course, this definition should be such that the corresponding integral is meaningful and has the desired properties. For example, if we choose

$$\tau_i = (1 - a) t_{i-1} + a t_i, \quad 0 \leqq a \leqq 1, i = 1, 2, \ldots, n,$$

we obtain from (4.2.3)

$$\underset{\delta_n \to 0}{\text{qm-lim}} S_n = (W_t^2 - W_{t_0}^2)/2 + (a - 1/2) (t - t_0)$$

(4.2.4)
$$= ((a)) \int_{t_0}^{t} W_s \, dW_s.$$

In particular, if we choose $a = 0$, that is, $\tau_i = t_{i-1}$, we obtain the **Itô**, or **stochastic integral**, with which we are concerned almost exclusively in the present book. From (4.2.4),

(4.2.5)   (Itô) $\int\limits_{t_0}^{t} W_s \, dW_s = ((0)) \int\limits_{t_0}^{t} W_s \, dW_s = (W_t^2 - W_{t_0}^2)/2 - (t - t_0)/2$.

Among integrals of the type (4.2.4), Itô's integral is characterized by the fact that, as a function of the upper limit, it is a martingale. This can be seen as follows: If, for simplicity, we take $t_0 = 0$ and

$$X_t = W_t^2/2 + (a - 1/2) \, t,$$

then, for $t \geqq s$, we have, with probability 1,

$$E(X_t | X_u, u \leqq s)$$

$$= E(W_t^2 | W_u^2/2 + (a - 1/2) \, u, u \leqq s)/2 + (a - 1/2) \, t$$

$$= E(E(W_t^2 | W_u, u \leqq s) | W_u^2/2 + (a - 1/2) \, u, u \leqq s)/2 + (a - 1/2) \, t$$

$$= E(E(W_t^2 | W_s) | W_u^2/2 + (a - 1/2) \, u, u \leqq s)/2 + (a - 1/2) \, t$$

$$= E(t - s + W_s^2 | W_u^2/2 + (a - 1/2) \, u, u \leqq s)/2 + (a - 1/2) \, t$$

$$= (t - s)/2 + (a - 1/2) \, t + W_s^2/2$$

$$= X_s + a \, (t - s),$$

where we have used the various simple properties of the conditional expectation and of a Wiener process. The process X_t *is therefore a martingale*, that is,

$$E(X_t | X_u, u \leqq s) = X_s \quad \text{with probability 1,}$$

if and only if $a = 0$, hence for Itô's choice of intermediate points. Similarly, we have $E \, X_t = a \, t \equiv 0$ if and only if $a = 0$.

It is disconcerting that (4.2.5) does not coincide with the value obtained in (4.2.1) by formal application of the classical rules. This disadvantage could be removed by the choice $a = 1/2$, but this would entail other and more serious disadvantages.

The applicability of the rules of classical Riemann-Stieltjes calculus was also the motivation for the definition of a stochastic integral given by R. L. Stratonovich [48]. In our special case,

$$\text{(Strat)} \int\limits_{t_0}^{t} W_s \, dW_s = \underset{\delta_n \to 0}{\text{qm-lim}} \sum_{i=1}^{n} \frac{W_{t_{i-1}} + W_{t_i}}{2} (W_{t_i} - W_{t_{i-1}})$$

$$= ((1/2)) \int\limits_{t_0}^{t} W_s \, dW_s = (W_t^2 - W_{t_0}^2)/2 .$$

We shall return to this in Chapter 10 and shall give a conversion formula.

4.3 Nonanticipating Functions

In the example considered in the preceding section, to every decomposition

$t_0 < t_1 < \ldots < t_n = t$ of the interval $[t_0, t]$ and every choice of intermediate points $\tau_i \in [t_{i-1}, t_i]$ there corresponds a step function

$$W_s^{(n)} = \begin{cases} \sum_{i=1}^n W_{\tau_i} I_{[t_{i-1}, t_i]}(s), & t_0 \leq s < t, \\ W_t, & s = t, \end{cases}$$

approximating the integrand W_s. If $\delta_n = \max (t_i - t_{i-1}) \to 0$, then, by virtue of the continuity of W_s,

$$ac\text{-}\lim W_s^{(n)} = W_s$$

uniformly in $[t_0, t]$, regardless of the choice of the intermediate points τ_i.

Although we define the integral of the approximating step function $W_s^{(n)}$ with respect to W_s as the corresponding Riemann-Stieltjes sum

$$\int_{t_0}^t W_s^{(n)} \, dW_s = S_n = \sum_{i=1}^n W_{\tau_i} (W_{t_i} - W_{t_{i-1}}),$$

the existence and value of the limit of the S_n depend on the intermediate points τ_i.

A particular property of Itô's choice of intermediate points, namely, $\tau_i = t_{i-1}$, is that the value of the corresponding approximating step function $W_s^{(n)}$ can, at every fixed instant $s \in [t_0, t]$, be obtained from knowledge of the values of W. from t_0 up to the instant s. (Actually, knowledge of $W_{t_{i-1}}$ is all we need to be able to determine $W_s^{(n)}$ for $s \in [t_{i-1}, t_i)$.) In other words, $W_s^{(n)}$ is $\mathfrak{W}[t_0, s]$-measurable, where

(4.3.1) $\mathfrak{W}[t_0, s] = \mathfrak{A}(W_u; t_0 \leq u \leq s)$.

We say that $W_s^{(n)}$ is a **nonanticipating function** of W. Itô's stochastic integral [42] has been defined for a broad class of these nonanticipating functions. Therefore, let us look at this concept.

Let W_t denote an m-dimensional Wiener process defined on the probability space $(\Omega, \mathfrak{A}, P)$. Suppose that t_0 is a fixed nonnegative number, that $\mathfrak{W}[t_0, t]$ is the sigma-algebra (4.3.1), and that

$$\mathfrak{W}_t^+ = \mathfrak{A}(W_s - W_t; t \leq s < \infty).$$

We recall the intuitive meaning of these two sub-sigma-algebras of $\mathfrak{A}$. Roughly speaking, they say that $\mathfrak{W}[t_0, t]$, for example, contains all those events that are defined by the conditions on the course of process W. in the interval $[t_0, t]$ (and nowhere else).

Since W_t has independent increments, $\mathfrak{W}[t_0, t]$ and $\mathfrak{W}_t^+$ are independent.

(4.3.2) **Definition.** Let t_0 denote a fixed nonnegative number. A family $\mathfrak{F}_t$, for $t \geq t_0$, of sub-sigma-algebras of $\mathfrak{A}$ is said to be **nonanticipating** with respect to the m-dimensional Wiener process W_t if it has the following three properties:

(a) $\mathfrak{F}_s \subset \mathfrak{F}_t \ (t_0 \leq s \leq t)$,

(b) $\mathfrak{F}_t \supset \mathfrak{W}\,[t_0, t] \ (t \geq t_0)$,

(c) $\mathfrak{F}_t$ is independent of $\mathfrak{W}_t^+ \ (t \geq t_0)$.

Since $\mathfrak{W}_0^+ = \mathfrak{W}\,[0, \infty)$ (apart from sets of measure 0), condition (c) means, for example, for $t = 0$, that $\mathfrak{F}_0$ can contain only events that are independent of the entire Wiener process W_t for $t \geq 0$.

(4.3.3) **Example.** The family

$$\mathfrak{F}_t = \mathfrak{W}\,[t_0, t]$$

is the smallest possible nonanticipating family of sigma-algebras. However, it is often necessary and desirable to augment $\mathfrak{W}\,[t_0, t]$ with other events that are independent of $\mathfrak{W}_t^+$ (for example, initial conditions). In the case of stochastic differential equations, we usually take

$$\mathfrak{F}_t = \mathfrak{A}\,(\mathfrak{W}\,[t_0, t], c),$$

where c is a random variable independent of $\mathfrak{W}_{t_0}^+$.

(4.3.4) **Definition.** A $(d \times m$ matrix)-valued function $G = G\,(s, \omega)$ defined on $[t_0, t] \times \Omega$ and measurable in (s, ω) is said to be **nonanticipating** (with respect to a family $\mathfrak{F}_s$ of nonanticipating sigma-algebras) if $G\,(s, .)$ is $\mathfrak{F}_s$-measurable for all $s \in [t_0, t]$. We denote by $M_2^{d, m}\,[t_0, t] = M_2\,[t_0, t]$ the set of those nonanticipating functions defined on $[t_0, t] \times \Omega$ for which the sample functions $G\,(., \omega)$ are with probability 1 in $L_2\,[t_0, t]$, that is, with probability 1

$$\int\limits_{t_0}^{t} |G\,(s, \omega)|^2 \ ds < \infty\,.$$

Here, the last integral is to be interpreted as the Lebesgue integral (which, for example, coincides with the Riemann integral in the case of continuous functions). We denote by

$$|G| = \left(\sum_{i=1}^{d} \sum_{j=1}^{m} G_{ij}^2\right)^{1/2} = (\mathrm{tr}\ G\,G')^{1/2}$$

the norm of the matrix G. We have $G \in M_2^{d, m}\,[t_0, t]$ if and only if $G_{ij} \in M_2^{1, 1}\,[t_0, t]$ for all i and j.

Furthermore,

$$M_2\,[t_0, s] \supset M_2\,[t_0, t], \quad t_0 < s \leq t.$$

We set

$$M_2 = M_2^{d, m} = \bigcap_{t > t_0} M_2 [t_0, t].$$

(4.3.5) **Example.** Every function $G(t, \omega) \equiv G(t)$ that is independent of ω is of course always nonanticipating. Such a function belongs to $M_2 [t_0, t]$ if and only if it belongs $L_2 [t_0, t]$.

(4.3.6) **Example.** Since $\mathfrak{F}_s \supset \mathfrak{W} [t_0, s]$, the function G is nonanticipating for every choice of $\mathfrak{F}_s$ if $G(s, .)$ is $\mathfrak{W} [t_0, s]$-measurable, that is,

$$G(s, \omega) = \bar{G}(s; W_r(\omega), t_0 \leqq r \leqq s), \quad t_0 \leqq s \leqq t.$$

In this case, $G(s, .)$ is thus a functional of the sample functions of W_r in the interval $[t_0, s]$. An example of this is the case $G(s, .) = W_s$, discussed in section 4.2. A more complicated example is

$$G(s, .) = \max_{t_0 \leqq r \leqq s} |W_r|.$$

(4.3.7) **Example.** For $\mathfrak{F}_s = \mathfrak{W} [t_0, s]$, the function

$$G(s, .) = \max_{t_0 \leqq r \leqq 2s} |W_r|$$

is *not* $\mathfrak{F}_s$-measurable; that is, G is anticipating.

(4.3.8) **Remark.** If G is nonanticipating, every measurable function $g(t, G)$ is nonanticipating. Furthermore, $M_2 [t_0, t]$ is a linear space.

4.4 Definition of the Stochastic Integral

The purpose of the present section is to define the stochastic integral

$$\int_{t_0}^{t} G \, dW = \int_{t_0}^{t} G(s) \, dW_s = \int_{t_0}^{t} G(s, \omega) \, dW_s(\omega)$$

for arbitrary $t \geqq t_0$ and all $G \in M_2^{d, m} [t_0, t] = M_2 [t_0, t]$. We shall do this in two steps. In the first step, we define the integral for step functions in $M_2 [t_0, t]$. In the second step, we extend this definition to the entire set $M_2 [t_0, t]$ by means of an approximation of an arbitrary function with the aid of step functions.

Step 1. A function $G \in M_2 [t_0, t]$ is called a **step function** if there exists a decomposition $t_0 < t_1 < \dots < t_n = t$ such that $G(s) = G(t_{i-1})$ (note that we omit the variable ω) for all $s \in [t_{i-1}, t_i)$, where $i = 1, \dots, n$. For such step functions, we define the **stochastic integral** of G with respect to W_t as the R^d-valued random variable

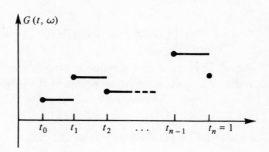

Fig. 5:
A nonanticipating step function.

(4.4.1) $\quad \int\limits_{t_0}^{t} G \, dW = \int\limits_{t_0}^{t} G(s) \, dW_s = \sum\limits_{i=1}^{n} G(t_{i-1}) (W_{t_i} - W_{t_{i-1}}).$

We summarize some of the more important properties of this integral in

(4.4.2) **Theorem.** Let t denote a number equal to or greater than t_0, let a and b denote members of R^1, and let G_1 and G_2 denote step functions belonging to $M_2 [t_0, t]$. Then, for the stochastic integral (4.4.1), we have

a) $\quad \int\limits_{t_0}^{t} (a \, G_1 + b \, G_2) \, dW = a \int\limits_{t_0}^{t} G_1 \, dW + b \int\limits_{t_0}^{t} G_2 \, dW.$

b) Also,

$$\int\limits_{t_0}^{t} G \, dW = \begin{pmatrix} \sum\limits_{k=1}^{m} \int\limits_{t_0}^{t} G_{1k}(s) \, dW_s^k \\ \vdots \\ \sum\limits_{k=1}^{m} \int\limits_{t_0}^{t} G_{dk}(s) \, dW_s^k \end{pmatrix}, \quad W_t = \begin{pmatrix} W_t^1 \\ \vdots \\ W_t^m \end{pmatrix}.$$

c) For $E |G(s)| < \infty$ for all s in $[t_0, t]$ we have

$$E \left(\int\limits_{t_0}^{t} G \, dW \right) = 0.$$

d) For $E |G(s)|^2 < \infty$, where $s \in [t_0, t]$, the following holds for the $d \times d$ covariance matrix of the stochastic integral (4.4.1):

(4.4.3) $\quad E \left(\int\limits_{t_0}^{t} G \, dW \right) \left(\int\limits_{t_0}^{t} G \, dW \right)' = \int\limits_{t_0}^{t} E \, G(s) \, G(s)' \, ds,$

and, in particular,

(4.4.4) $E \left| \int_{t_0}^{t} G \, dW \right|^2 = \int_{t_0}^{t} E \, |G|^2 \, ds.$

Proof. a) A linear combination of step functions is a step function with break points the same as the break points of the original functions (or possibly a portion of them). The asserted linearity of the integral follows immediately from the definition (4.4.1).

b) also follows immediately from the definition.

c) From (4.4.1), we have

$$E \left(\int_{t_0}^{t} G \, dW \right) = \sum_{i=1}^{n} E \, G \, (t_{i-1}) \, E \, (W_{t_i} - W_{t_{i-1}})$$

since $G \, (t_{i-1})$ is independent of $W_{t_i} - W_{t_{i-1}}$. Since $E \, (W_{t_i} - W_{t_{i-1}}) = 0$, the assertion follows.

d) From (4.4.1), we have

$$E \left(\int_{t_0}^{t} G \, dW \right) \left(\int_{t_0}^{t} G \, dW \right)'$$

$$= \sum_{i=1}^{n} \sum_{j=1}^{n} E \, [G \, (t_{i-1}) \, (W_{t_i} - W_{t_{i-1}}) \, (W_{t_j} - W_{t_{j-1}})' \, G \, (t_{j-1})']$$

$$= \sum_{i=1}^{n} E \, [G \, (t_{i-1}) \, (W_{t_i} - W_{t_{i-1}}) \, (W_{t_i} - W_{t_{i-1}})' \, G \, (t_{i-1})']$$

$$+ 2 \sum_{i<j} E \, [G \, (t_{i-1}) \, (W_{t_i} - W_{t_{i-1}}) \, (W_{t_j} - W_{t_{j-1}})' \, G \, (t_{j-1})']$$

$$= S_1 + S_2.$$

In particular, the matrix element c_{kh}^i in the ith summand of S_1 is

$$c_{kh}^i = \sum_{p=1}^{m} \sum_{q=1}^{m} E \, G_{kp} \, (t_{i-1}) \, (W_{t_i}^p - W_{t_{i-1}}^p) \, (W_{t_i}^q - W_{t_{i-1}}^q) \, G_{hq} \, (t_{i-1})$$

$$= \sum_{p=1}^{m} \sum_{q=1}^{m} E \, (G_{kp} \, (t_{i-1}) G_{hq} \, (t_{i-1})) \, E ((W_{t_i}^p - W_{t_{i-1}}^p)(W_{t_i}^q - W_{t_{i-1}}^q)),$$

where we have used the fact that G is nonanticipating. Now, $W_t - W_s$ has the distribution $\mathfrak{N} \, (0, |t - s| \, I)$, so that

$$c_{kh}^i = \sum_{p=1}^{m} E \, (G_{kp} \, (t_{i-1}) \, G_{hp} \, (t_{i-1})) \, (t_i - t_{i-1})$$

$$= E \, (G \, (t_{i-1}) \, G \, (t_{i-1})')_{kh} \, (t_i - t_{i-1}) \, .$$

Therefore,

$$S_1 = \sum_{i=1}^{n} E\left(G\left(t_{i-1}\right) G\left(t_{i-1}\right)'\right)\left(t_i - t_{i-1}\right) = \int_{t_0}^{t} E\, G\left(s\right) G\left(s\right)'\, ds.$$

We now treat S_2 in the same way and again use the fact that the terms $G_{kp}\left(t_{i-1}\right)\left(W_{t_i}^p - W_{t_{i-1}}^p\right) G_{hq}\left(t_{j-1}\right)$ and $W_{t_j}^q - W_{t_{j-1}}^q$ are, for $i < j$, independent. This yields

$$S_2 = 0,$$

the desired outcome. Since tr $(G\, G') = |G|^2$, equation (4.4.4) follows from (4.4.3) by applying the trace operator. ∎

One should note that we do not assume that $E\, G\left(s\right) = 0$ in Theorem (4.4.2c). Rather, the stochastic integral (4.4.1) has expectation 0 *in every case*. Also, the form of the covariance matrix (4.4.3) is amazingly simple.

Step 2:   Definition of the stochastic integral for arbitrary functions in $M_2\left[t_0, t\right]$:

Let us show first that the set of step functions is dense in $M_2\left[t_0, t\right]$ in the sense of the following lemma:

(4.4.5)   **Lemma.**   For every function $G \in M_2\left[t_0, t\right]$, there exists a series of step functions $G_n \in M_2\left[t_0, t\right]$ such that

$$\underset{n \to \infty}{\text{ac-lim}} \int_{t_0}^{t} |G\left(s\right) - G_n\left(s\right)|^2\, ds = 0.$$

Proof. If $G\left(.,\omega\right)$ is continuous with probability 1, we can approximate that function uniformly and hence in the sense of mean square with the nonanticipating step functions

$$G_n\left(s\right) = G\left(t_0 + k\left(t - t_0\right)/n\right),$$

$$t_0 + k\left(t - t_0\right)/n \le s < t_0 + \left(k + 1\right)\left(t - t_0\right)/n, \quad 0 \le k \le n.$$

If G is bounded by a constant c independent of s and ω, there exists a sequence $\{G_n\}$ of continuous functions in $M_2\left[t_0, t\right]$, for example

$$G_n\left(s\right) = n \int_{t_0}^{s} e^{n\left(u-s\right)} G\left(u\right)\, du,$$

that is uniformly bounded by the constant c and converges, for almost all $s \in \left[t_0, t\right]$ $\left[\lambda\right]$, to G—all this with probability 1. If we apply the theorem of dominated convergence to the variable s (see section 1.3), it then follows with probability 1 that G_n converges to G in the sense of $L_2\left[t_0, t\right]$.

Finally, if G is any function in $M_2\left[t_0, t\right]$, we can, for example, by shifting to $G\left(t, \omega\right) I_{[|G\left(t, \omega\right)| \le c]}$, approximate it to an arbitrary degree of accuracy in the sense of $L_2 [t_0, t]$ with a function in $M_2 [t_0, t]$ whose sample functions are bounded by a constant c .

Thus, the set of step functions is dense in the sense of convergence in $M_2 [t_0, t]$. ∎

If the assertion of Lemma (4.4.5) holds for a function $G \in M_2 [t_0, t]$ and a sequence $\{G_n\}$ of step functions, the weaker assertion

$$\text{st-}\lim_{n \to \infty} \int_{t_0}^{t} |G(s) - G_n(s)|^2 \, ds = 0$$

of course follows (see section 1.4). We now wish to show that this last assertion implies stochastic convergence of the sequence of integrals

$$\int_{t_0}^{t} G_n(s) \, dW_s$$

to a specific random variable. For this we shall use the following estimate for the stochastic integral of step functions.

(4.4.6) **Lemma.** Suppose that $G \in M_2 [t_0, t]$ is a step function. Then, for all $N > 0$ and $c > 0$,

$$P \left[\left| \int_{t_0}^{t} G(s) \, dW_s \right| > c \right] \leq N/c^2 + P \left[\int_{t_0}^{t} |G(s)|^2 \, ds > N \right].$$

Proof. Suppose that $G(s) = G(t_{i-1})$ for $t_{i-1} \leq s < t_i$, where $t_0 < t_1 < \ldots < t_n = t$. The function

$$G_N(s) = \begin{cases} G(s) & \text{if } \int_{t_0}^{t_i} |G(s)|^2 \, ds \leq N, \, t_{i-1} \leq s < t_i, \\[2em] 0 & \text{if } \int_{t_0}^{t_i} |G(s)|^2 \, ds > N, \, t_{i-1} \leq s < t_i, \end{cases}$$

is a nonanticipating step function and hence belongs to $M_2 [t_0, t]$ since

$$\int_{t_0}^{t_i} |G(s)|^2 \, ds \quad \text{is} \quad \mathfrak{F}_{t_{i-1}}\text{-measurable.}$$

Since

$$\int_{t_0}^{t} |G_N(s)|^2 \, ds = \sum_{i=1}^{n} |G_N(t_{i-1})|^2 \, (t_i - t_{i-1}) \leq N,$$

we have $|G_N(t_{i-1})|^2 \leq N/(t_i - t_{i-1})$, so that

$$E\,|G_N(s)|^2 < \infty.$$

From Theorem (4.4.2d), we therefore have

$$(4.4.7) \qquad E\left|\int_{t_0}^{t} G_N(s)\,dW_s\right|^2 = \int_{t_0}^{t} E\,|G_N(s)|^2\,ds \leq N.$$

Finally, $G_N \neq G$ if and only if

$$\int_{t_0}^{t} |G(s)|^2\,ds > N,$$

so that

$$(4.4.8) \qquad P[\sup_{t_0 \leq s \leq t} |G_N(s) - G(s)| > 0] = P\left[\int_{t_0}^{t} |G(s)|^2\,ds > N\right].$$

If we use equations (4.4.7) and (4.4.8), Theorem (4.4.2a), and the triangle and Chebyshev inequalities, we obtain

$$P\left[\left|\int_{t_0}^{t} G(s)\,dW_s\right| > c\right]$$

$$\leq P\left[\left|\int_{t_0}^{t} G_N(s)\,dW_s\right| > c\right] + P\left[\left|\int_{t_0}^{t} (G(s) - G_N(s))\,dW_s\right| > 0\right]$$

$$\leq E\left|\int_{t_0}^{t} G_N(s)\,dW_s\right|^2 \Big/ c^2 + P\left[\int_{t_0}^{t} |G(s)|^2\,ds > N\right]$$

$$\leq N/c^2 + P\left[\int_{t_0}^{t} |G(s)|^2\,ds > N\right]. \quad \blacksquare$$

(4.4.9) **Lemma.** Suppose that $G \in M_2[t_0, t]$ and that $G_n \in M_2[t_0, t]$ is a sequence of step functions for which

$$(4.4.10) \qquad \text{st-}\lim_{n \to \infty} \int_{t_0}^{t} |G(s) - G_n(s)|^2\,ds = 0.$$

If we define

$$\int_{t_0}^{t} G_n(s)\,dW_s$$

by equation (4.4.1), then

$$\text{st-}\lim_{n \to \infty} \int_{t_0}^{t} G_n(s)\, dW_s = I(G),$$

where $I(G)$ is a random variable that does not depend on the special choice of the sequence $\{G_n\}$.

Proof. Since

$$\int_{t_0}^{t} |G_n - G_m|^2\, ds \leqq 2 \int_{t_0}^{t} |G - G_n|^2\, ds + 2 \int_{t_0}^{t} |G - G_m|^2\, ds,$$

it follows from the assumption that

$$\text{st-}\lim \int_{t_0}^{t} |G_n - G_m|^2\, ds = 0$$

as $n, m \to \infty$. This is the same as saying

$$\lim_{n, m \to \infty} P\left[\int_{t_0}^{t} |G_n(s) - G_m(s)|^2\, ds > \varepsilon \right] = 0 \text{ for all } \varepsilon > 0.$$

Therefore, application of Lemma (4.4.6) to $G_n - G_m$ yields

$$\limsup_{n, m \to \infty} P\left[\left| \int_{t_0}^{t} G_n(s)\, dW_s - \int_{t_0}^{t} G_m(s)\, dW_s \right| > \delta \right] \leqq \varepsilon/\delta^2$$

$$+ \limsup_{n, m \to \infty} P\left[\int_{t_0}^{t} |G_n(s) - G_m(s)|^2\, ds > \varepsilon \right]$$

$$= \varepsilon/\delta^2.$$

Since ε is an arbitrary positive number, we have

$$\lim_{n, m \to \infty} P\left[\left| \int_{t_0}^{t} G_n(s)\, dW_s - \int_{t_0}^{t} G_m(s)\, dW_s \right| > \delta \right] = 0.$$

Since every stochastic Cauchy sequence also converges stochastically, there exists a random variable $I(G)$ such that

$$\int_{t_0}^{t} G_n(s)\, dW_s \to I(G) \quad \text{(stochastically)}.$$

The limit is almost certainly uniquely determined and independent of the special choice of sequence $\{G_n\}$ for which (4.4.10) holds. This is true because, if $\{G_n\}$ and $\{\overline{G}_n\}$ are two such sequences, we can combine them into a single sequence, from which the almost certain coincidence of the corresponding limits follows. ∎

The following definition follows from Lemma (4.4.9).

(4.4.11) **Definition.** For every ($d \times m$ matrix)-valued function $G \in M_2[t_0, t]$, the **stochastic integral** (or **Itô's integral**) of G with respect to the m-dimensional Wiener process W_t over the interval $[t_0, t]$ is defined as the random variable $I(G)$, which is almost certainly uniquely determined in accordance with Lemma (4.4.9):

$$\int_{t_0}^{t} G \, dW = \int_{t_0}^{t} G(s) \, dW_s = \underset{n \to \infty}{\text{st-lim}} \int_{t_0}^{t} G_n \, dW,$$

where $\{G_n\}$ is a sequence of step functions in $M_2[t_0, t]$ that approximates G in the sense of

$$\underset{n \to \infty}{\text{st-lim}} \int_{t_0}^{t} |G(s) - G_n(s)|^2 \, ds = 0.$$

For special functions in $M_2[t_0, t]$, we can give a stronger than mere stochastic approximation of the stochastic integral. Specifically, we can approximate it in mean square, as indicated by the following lemma:

(4.4.12) **Lemma.** For every function $G \in M_2[t_0, t]$ such that

(4.4.13) $\quad \int_{t_0}^{t} E |G(s)|^2 \, ds < \infty,$

there exists a sequence $\{G_n\}$ of step functions in $M_2[t_0, t]$ with the same property, so that

$$\lim_{n \to \infty} \int_{t_0}^{t} E |G_n(s) - G(s)|^2 \, ds = 0$$

and

$$\underset{n \to \infty}{\text{qm-lim}} \int_{t_0}^{t} G_n(s) \, dW_s = \int_{t_0}^{t} G(s) \, dW_s.$$

Proof. According to Lemma (4.4.9), there always exists a sequence $\{\overline{G}_n\}$ of step functions such that

$$\text{st-}\lim_{} \int_{t_0}^{t} |G - \overline{G}_n|^2 \, ds = 0.$$

Let

$$g_N(x) = \begin{cases} x, & |x| \leqq N, \\ N x/|x|, & |x| > N. \end{cases}$$

Since $|g_N(x) - g_N(y)| \leqq |x - y|$, it follows that

$$\int_{t_0}^{t} |g_N(G(s)) - g_N(\overline{G}_n(s))|^2 \, ds \leqq \int_{t_0}^{t} |G(s) - \overline{G}_n(s)|^2 \, ds \rightarrow 0 \text{ (stochastically)}.$$

Since

$$\int_{t_0}^{t} |g_N(G(s)) - g_N(\overline{G}_n(s))|^2 \, ds \leqq 4 N^2 t ,$$

it follows from the theorem on dominated convergence (with respect to the variable ω) that

$$E \int_{t_0}^{t} |g_N(G) - g_N(\overline{G}_n)|^2 \, ds = \int_{t_0}^{t} E |g_N(G) - g_N(\overline{G}_n)|^2 \, ds \rightarrow 0$$

as $n \rightarrow \infty$. It follows from the same theorem (now applied to the variable $(s, \omega) \in [t_0, t] \times \Omega$) that

$$\int_{t_0}^{t} E |g_N(G(s)) - G(s)|^2 \, ds \rightarrow 0,$$

as $N \rightarrow \infty$ (by virtue of the inequality $|g_N(G(s)) - G(s)|^2 \leqq |G(s)|^2$ and the assumption (4.4.13)). Therefore, there exist sequences $\{N_k\}$ and $\{n_k\}$ such that

$$\int_{t_0}^{t} E |G(s) - g_{N_k}(\overline{G}_{n_k}(s))|^2 \, ds \leqq 2 \int_{t_0}^{t} E |G(s) - g_{N_k}(G(s))|^2 \, ds$$

$$+ 2 \int_{t_0}^{t} E |g_{N_k}(G(s)) - g_{N_k}(\overline{G}_{n_k}(s))|^2 \, ds$$

$$\rightarrow 0 \quad (k \rightarrow \infty).$$

Accordingly, we can choose

$$G_k(s) = g_{N_k}(\overline{G}_{n_k}(s))$$

so as to get a sequence of step functions such that

$$\lim_{k \to \infty} \int_{t_0}^{t} E \,|G_k(s) - G(s)|^2 \, ds = 0 \,.$$

For this sequence, we have

$$\underset{k \to \infty}{\text{st-lim}} \int_{t_0}^{t} G_k \, dW = \int_{t_0}^{t} G \, dW \,.$$

However, by virtue of Theorem (4.4.2d) and what we have just proven, we see that

$$E \left| \int_{t_0}^{t} G_k \, dW - \int_{t_0}^{t} G_p \, dW \right|^2 = \int_{t_0}^{t} E \,|G_k - G_p|^2 \, ds \longrightarrow 0$$

as $k, p \longrightarrow \infty$ and hence

$$\underset{k \to \infty}{\text{qm-lim}} \int_{t_0}^{t} G_k \, dW = \int_{t_0}^{t} G \, dW \,. \blacksquare$$

In extending the concept of an integral from step functions to arbitrary functions in $M_2 [t_0, t]$, the most important properties are naturally carried over also. We summarize these in

(4.4.14)   **Theorem.** Let $G, G_1, G_2,$ and G_n denote ($d \times m$ matrix)-valued functions in $M_2 [t_0, t]$ and let W_t denote an m-dimensional Wiener process. Then, the stochastic integral defined by (4.4.11) has the following properties:

a) $$\int_{t_0}^{t} (a \, G_1 + b \, G_2) \, dW = a \int_{t_0}^{t} G_1 \, dW + b \int_{t_0}^{t} G_2 \, dW, \quad a, b \in R^1 \,.$$

b) $$\int_{t_0}^{t} G \, dW = \sum_{k=1}^{m} \begin{pmatrix} \int_{t_0}^{t} G_{1k} \, dW^k \\ \vdots \\ \int_{t_0}^{t} G_{dk} \, dW^k \end{pmatrix}, \quad W_t = \begin{pmatrix} W_t^1 \\ \vdots \\ W_t^m \end{pmatrix}.$$

c) For $N > 0$ and $c > 0$,

$$P \left[\left| \int_{t_0}^{t} G \, dW \right| > c \right] \leq N/c^2 + P \left[\int_{t_0}^{t} |G|^2 \, ds > N \right].$$

d) The relationship

$$\operatorname*{st-lim}_{n \to \infty} \int_{t_0}^{t} |G(s) - G_n(s)|^2 \, ds = 0$$

implies

$$\operatorname*{st-lim}_{n \to \infty} \int_{t_0}^{t} G_n \, dW = \int_{t_0}^{t} G \, dW$$

(where the G_n need not be step functions).

e) If

$$\int_{t_0}^{t} E |G(s)|^2 \, ds < \infty,$$

then, for the expectation vector of the stochastic integral and its covariance matrix, we always have respectively

$$E \left(\int_{t_0}^{t} G \, dW \right) = 0$$

and

$$E \left(\int_{t_0}^{t} G \, dW \right) \left(\int_{t_0}^{t} G \, dW \right)' = \int_{t_0}^{t} E G G' \, ds;$$

hence, in particular,

$$E \left| \int_{t_0}^{t} G \, dW \right|^2 = \int_{t_0}^{t} E |G|^2 \, ds.$$

Proof. Parts a) and b) follow immediately from parts a) and b) of Theorem (4.4.2) by taking the limit. Similarly, part c) follows from Lemma (4.4.6). Part d) can now be obtained from part c) just as in the proof of Lemma (4.4.9).

For the proof of part e), we use the fact that, in accordance with Lemma (4.4.12), there always exists a sequence of step functions $\{G_n\}$ such that

$$\lim_{n \to \infty} \int_{t_0}^{t} E |G_n - G|^2 \, ds = 0$$

and

$$\operatorname*{qm-lim}_{n \to \infty} \int_{t_0}^{t} G_n \, dW = \int_{t_0}^{t} G \, dW .$$

Now using Theorem (4.4.2), parts c) and d), we get from the last equation

$$E \left(\int_{t_0}^{t} G_n \, dW \right) = 0 \to E \left(\int_{t_0}^{t} G \, dW \right) = 0$$

and

$$E \left(\int_{t_0}^{t} G_n \, dW \right) \left(\int_{t_0}^{t} G_n \, dW \right)' = \int_{t_0}^{t} E \, G_n \, G_n' \, ds$$

$$\to \int_{t_0}^{t} E \, G \, G' \, ds = E \left(\int_{t_0}^{t} G \, dW \right) \left(\int_{t_0}^{t} G \, dW \right)' . \quad \blacksquare$$

Actual evaluation of the stochastic integrals is another matter. Of course, this problem exists even with ordinary integrals. It is always possible to use the definition, which is of a constructive nature, to obtain *an arbitrarily* close approximation of any stochastic integral. Itô's theorem, which will be discussed in section 5.3, is an important tool for explicit evaluation of many stochastic integrals.

4.5 Examples and Remarks

The integral

(4.5.1) $X_t = \int_{t_0}^{t} W_s \, dW_s$,

which we discussed in section 4.2, can now be evaluated without difficulty, for example, in accordance with

(4.5.2) **Corollary.** Suppose that $G \in M_2 [t_0, t]$ is continuous with probability 1. Then,

$$\int_{t_0}^{t} G \, dW = \operatorname*{st-lim}_{\delta_n \to 0} \sum_{k=1}^{n} G \left(t_{k-1} \right) \left(W_{t_k} - W_{t_{k-1}} \right),$$

$$t_0 < t_1 < \ldots < t_n = t, \quad \delta_n = \max \left(t_k - t_{k-1} \right).$$

Proof. We have

$$\int\limits_{t_0}^{t} G_n \, \mathrm{d}W = \sum_{k=1}^{n} G\left(t_{k-1}\right)\left(W_{t_k} - W_{t_{k-1}}\right)$$

for the nonanticipating step functions

(4.5.3) $$G_n(s) = \sum_{k=1}^{n} G\left(t_{k-1}\right) I_{[t_{k-1}, t_k)}(s).$$

Therefore, in accordance with Lemma (4.4.9), what we need to show is that

$$\text{st-}\lim_{n \to \infty} \int\limits_{t_0}^{t} |G(s) - G_n(s)|^2 \, \mathrm{d}s = 0.$$

This is true by virtue of the continuity of $G(\cdot, \omega)$ with probability 1. ∎

Corollary (4.5.2) tells us that, for almost certainly continuous G, the simplest nonanticipating step functions of the form (4.5.3), namely, those obtained from the function itself, can be used for approximation of the stochastic integral. This corollary remains valid if we replace almost certain continuity with stochastic continuity.

If we apply this corollary to (4.5.1), we obtain

$$\int\limits_{t_0}^{t} W \, \mathrm{d}W = \text{st-}\lim_{\delta_n \to 0} \sum_{k=1}^{n} W_{t_{k-1}} \left(W_{t_k} - W_{t_{k-1}}\right)$$

$$= (W_t^2 - W_{t_0}^2)/2 - (t - t_0)/2,$$

which we can get from (4.2.4), for example.

For an m-dimensional Wiener process, we similarly obtain

$$\int\limits_{t_0}^{t} W_s' \, \mathrm{d}W_s = (|W_t|^2 - |W_{t_0}|^2)/2 - m\,(t - t_0)/2.$$

(4.5.4) Corollary. If

$$\int\limits_{t_0}^{t} |G|^2 \, \mathrm{d}s = 0 \quad \text{with probability 1,}$$

where $G \in M_2\,[t_0, t]$, then

$$\int\limits_{t_0}^{t} G \, \mathrm{d}W = 0 \quad \text{with probability 1.}$$

Proof. The sequence of step functions $G_n \equiv 0$ approximates G and yields the result. ∎

This last corollary tells us, for example, that the value of every function G can be changed for every s in a fixed set of Lebesgue measure 0 in $[t_0, t]$ without changing the value of its stochastic integral. In particular, this is true for the values of the function at a finite or countable set of points s.

(4.5.5) **Corollary.** If

$$\int_{t_0}^{t} E |G(s)|^2 \, ds < \infty,$$

where $G \in M_2 [t_0, t]$, then, for every positive c,

$$P\left[\left|\int_{t_0}^{t} G \, dW\right| > c\right] \le \int_{t_0}^{t} E |G(s)|^2 \, ds/c^2.$$

This is simply Chebyshev's inequality. It holds in particular for every function $G \in L_2 [t_0, t]$ that is independent of ω. However, in this case we can determine exactly the distribution of the stochastic intergral.

(4.5.6) **Corollary.** If G is independent of ω and belongs to $L_2 [t_0, t]$, it belongs to $M_2 [t_0, t]$ for any sub-sigma-algebras $\mathfrak{F}_s \supset \mathfrak{W} [t_0, s]$, and the stochastic integral

$$\int_{t_0}^{t} G \, dW$$

is a normally distributed d-dimensional random variable with distribution

$$\mathfrak{N}\left(0, \int_{t_0}^{t} G(s) \, G(s)' \, ds\right).$$

Proof. If G is independent of ω, we can find a sequence $\{G_n\}$ of step functions that are also independent of ω such that

(4.5.7) $$\int_{t_0}^{t} |G - G_n|^2 \, ds \longrightarrow 0$$

and (in accordance with (4.4.4))

(4.5.8) $$\text{qm-}\lim \int_{t_0}^{t} G_n \, dW = \int_{t_0}^{t} G \, dW.$$

Now the random variable

$$\int_{t_0}^{t} G_n \, dW = \sum_{i=1}^{n} G_n (t_{i-1}) (W_{t_i} - W_{t_{i-1}})$$

certainly has distribution $\Re \left(0, \int_{t_0}^{t} G_n\, G_n'\, ds \right)$. Since, by virtue of (4.5.7),

$$\int_{t_0}^{t} G_n\, G_n'\, ds \rightarrow \int_{t_0}^{t} G\, G'\, ds\,,$$

it follows that the limit in (4.5.8) is also normally distributed with the given first and second moments. ∎

(4.5.9) **Remark.** Even when the step functions $\{G_n\}$ are such that

$$\operatorname*{ac-lim}_{n \to \infty}\ \int_{t_0}^{t} |G_n(s) - G(s)|^2\, ds = 0,$$

this does *not* in general imply almost certain convergence of the integrals $\int_{t_0}^{t} G_n\, dW$ to $\int_{t_0}^{t} G\, dW$.

Let us now show that formula (4.1.6), which we have used as a definition of a stochastic integral for smooth functions G that are independent of ω, can be proven for considerably more general Itô stochastic integrals. In other words, Itô's integral and the stochastic integral (4.1.6) are consistent and coincide when the latter is defined. Somewhat more generally, we have

(4.5.10) **Corollary.** Suppose that $G \in M_2^{d,\,m}\,[t_0,\, t]$ and that the variation of $G\,(\cdot,\, \omega)$ defined on $[t_0,\, t]$ is almost certainly bounded. Then,

$$(4.5.11) \qquad \int_{t_0}^{t} G(s)\, dW_s = G(t)\, W_t - G(t_0)\, W_{t_0} - \left(\int_{t_0}^{t} W_s'\, dG(s)' \right)',$$

where the last integral is the usual Riemann-Stieltjes integral. If, in fact, $G\,(\cdot,\, \omega)$ is almost certainly continuously differentiable on $[t_0,\, t]$ (or, more generally, absolutely continuous) with derivative $\dot{G}$, then

$$(4.5.12) \qquad \int_{t_0}^{t} G(s)\, dW_s = G(t)\, W_t - G(t_0)\, W_{t_0} - \int_{t_0}^{t} \dot{G}(s)\, W_s\, ds.$$

Proof. By virtue of the continuity of $W_\cdot$, both integrals in (4.5.11) exist under our assumptions as ordinary Riemann-Stieltjes integrals, and (4.5.11) is the usual rule for integration by parts, which, for continuously differentiable G, takes the form (4.5.12). Of course, the stochastic integral of G with respect to $W_\cdot$ coincides with the Riemann-Stieltjes integral if the latter exists. ∎

We note that G in Corollary (4.5.1) is not necessarily independent of ω.

Chapter 5

The Stochastic Integral as a Stochastic Process, Stochastic Differentials

5.1 The Stochastic Integral as a Function of the Upper Limit

Let W_t again denote an m-dimensional Wiener process. Let t_0 denote a fixed non-negative number. Let $\{\mathfrak{F}_t; t \geq t_0\}$ denote a family of nonanticipating sigma-algebras. Let $M_2[t_0, T]$ denote the set of nonanticipating ($d \times m$ matrix)-valued functions (see (4.3.4)) for which we have defined the R^d-valued stochastic integral

$$\int_{t_0}^{T} G \, dW = \int_{t_0}^{T} G(s, \omega) \, dW_s(\omega).$$

Suppose that G belongs to $M_2[t_0, T]$, that $A \subset [t_0, T]$ is a Borel set, and that I_A is its indicator function. Then,

$$G I_A \in M_2[t_0, T].$$

We therefore define

$$\int_{A} G \, dW = \int_{t_0}^{T} G I_A \, dW.$$

In accordance with Theorem (4.4.14a), for any two *disjoint* sets $A, B \subset [t_0, T]$, we have

$$\int_{A \cup B} G \, dW = \int_{A} G \, dW + \int_{B} G \, dW.$$

In particular, for $t_0 \leq a \leq b \leq c \leq T$ (by virtue of Corollary (4.5.4), finitely many points do not change the situation),

$$\int_{a}^{c} G \, dW = \int_{a}^{b} G \, dW + \int_{b}^{c} G \, dW.$$

In particular, for every G in $M_2[t_0, T]$,

$$X_t = \int_{t_0}^{t} G(s)\, dW_s = \int_{t_0}^{T} G(s)\, I_{[0,t]}\, dW_s$$

is an R^d-valued stochastic process defined (uniquely up to stochastic equivalence) for all $t \in [t_0, T]$ such that

$$X_{t_0} = 0 \quad \text{almost certainly.}$$

We have

$$X_t - X_s = \int_{s}^{t} G(u)\, dW_u, \quad t_0 \leqq s \leqq t \leqq T.$$

Now if G belongs to M_2, that is, to $M_2[t_0, t]$ for all $t \geqq t_0$, then X_t is defined for all $t \geqq t_0$. Then, all the assertions made in this chapter regarding X_t are valid (if the corresponding assumptions are satisfied) without any upper bound on the time, that is, for arbitrarily large intervals $[t_0, T]$.

We now wish to investigate the process X_t for fixed $G \in M_2[t_0, T]$. For this, we shall always assume that we have chosen a *separable* version of X_t (see section 1.8), which is always possible.

(5.1.1)   **Theorem.** Let G denote a function in $M_2[t_0, T]$ and suppose that

$$X_t = \int_{t_0}^{t} G(s)\, dW_s, \quad t_0 \leqq t \leqq T.$$

Then, the following hold:

a) X_t is $\mathfrak{F}_t$-measurable (and hence nonanticipating).

b) If

$$(5.1.2) \qquad \int_{t_0}^{t} E\,|G(s)|^2\, ds < \infty \quad \text{for all } t \leqq T,$$

then $(X_t, \mathfrak{F}_t)$, for $t \in [t_0, T]$, is an R^d-valued martingale; that is, for $t_0 \leqq s \leqq t \leqq T$,

$$E(X_t | \mathfrak{F}_s) = X_s.$$

Furthermore, for $t, s \in [t_0, T]$ we have

$$E\,X_t = 0,$$

(5.1.3) $\quad E\,X_t\,X_s' = \int\limits_{t_0}^{\min\,(t,\,s)} E\,G\,(u)\,G\,(u)'\,du\,;$

in particular,

(5.1.3a) $\quad E\,|X_t|^2 = \int\limits_{t_0}^{t} E\,|G\,(u)|^2\,du$

and, for all $c > 0$ and $t_0 \leq a \leq b \leq T$,

(5.1.4) $\quad P\,[\,\sup_{a \leq t \leq b}\,|X_t - X_a| > c\,] \leq \int\limits_{a}^{b} E\,|G\,(s)|^2\,ds/c^2$

and

(5.1.5) $\quad E\,(\,\sup_{a \leq t \leq b}\,|X_t - X_a|^2) \leq 4 \int\limits_{a}^{b} E\,|G\,(s)|^2\,ds.$

c) X_t has *continuous* sample functions with probability 1.

d) If, for some natural number k,

$$\int\limits_{a}^{t} E\,|G\,(s)|^{2k}\,ds < \infty, \quad t_0 \leq a \leq t \leq T,$$

then

(5.1.6) $\quad E\,|X_t - X_a|^{2k} \leq (k\,(2\,k-1))^{k-1}\,(t-a)^{k-1} \int\limits_{a}^{t} E\,|G\,(s)|^{2k}\,ds.$

Proof. a) The $\mathfrak{F}_t$-measurability of X_t follows from the definition

$$X_t^{(n)} = \int\limits_{t_0}^{t} G_n\,dW \;\rightarrow\; \int\limits_{t_0}^{t} G\,dW = X_t \quad \text{(stochastically)}$$

and the obvious $\mathfrak{F}_t$-measurability of the integral $X_t^{(n)}$ of the step functions G_n.

b) $\mathfrak{F}_t$ is an increasing family of sub-sigma-algebras $\mathfrak{A}$. By part a) of the present theorem, X_t is $\mathfrak{F}_t$-measurable. By Theorem (4.4.14e), X_t possesses, under the assumption (5.1.2), finite first and second moments. Thus, we need to show (see section 1.9) that, for $t_0 \leq s \leq t \leq T$,

$$E\,(X_t|\mathfrak{F}_s) = X_s$$

or, equivalently,

$$E\left(X_t - X_s \mid \mathfrak{F}_s\right) = E\left(\int_s^t G\,dW \mid \mathfrak{F}_s\right) = 0.$$

This certainly holds for step functions since $\mathfrak{F}_s$ and $\mathfrak{W}_s^+$ are independent and $E\left(W_{t_i} - W_{t_{i-1}}\right) = 0$ and hence in general. Similarly, formula (5.1.3) for the covariance matrix of the process X_t is first proven easily for step functions and then in general by taking the limit.

We know now that $(X_t, \mathfrak{F}_t)$ is a martingale. Hence, $(X_t - X_a, \mathfrak{F}_t)$, for $t \geq a$, is also a martingale and $(|X_t - X_a|^2, \mathfrak{F}_t)$ is a submartingale (see section 1.9). Then, inequality (1.9.1) yields, for every $c > 0$, $t_0 \leq a \leq b \leq T$, and $p = 2$,

(5.1.7) $P\left[\sup_{a \leq t \leq b} |X_t - X_a| > c\right] \leq \dfrac{1}{c^2}\, E\,|X_b - X_a|^2.$

From (5.1.3), we obtain

$$E\,|X_t - X_a|^2 = E\,X_t'\,X_t - E\,X_t'\,X_a - E\,X_a'\,X_t + E\,X_a'\,X_a = \int_a^t E\,|G(s)|^2\,ds,$$

so that

$$E\,|X_b - X_a|^2 = \int_a^b E\,|G(s)|^2\,ds.$$

If we substitute this into (5.1.7), we obtain (5.1.4). The estimate (5.1.5) follows from inequality (1.9.2) for $p = 2$.

c) The continuity of X_t is proven in three steps. Let T denote any finite number.

Step 1.   If G is a step function defined on $[t_0, T]$, then

$$X_t = \int_{t_0}^T G\,I_{[0,\,t]}\,dW = \sum_{t_i \leq t} G(t_{i-1})(W_{t_i} - W_{t_{i-1}}) + G(\max_{t_i \leq t} t_i)$$
$$(W_t - W_{\max_{t_i \leq t} t_i}).$$

The continuity of X_t follows from this formula and the continuity of W_t.

Step 2.   If G is a function in $M_2\,[t_0, T]$ such that

$$\int_{t_0}^T E\,|G|^2\,ds < \infty,$$

we choose on the basis of Lemma (4.4.12) a sequence of step functions G_n such that

$$\lim_{n \to \infty} \int_{t_0}^{T} E\,|G\,(s) - G_n\,(s)|^2\,\mathrm{d}s = 0.$$

By virtue of (5.1.4), if we set

$$X_t^{(n)} = \int_{t_0}^{t} G_n\,\mathrm{d}W,$$

we obtain

$$P\,[\sup_{t_0 \le t \le T} |X_t - X_t^{(n)}| > c] \le \int_{t_0}^{T} E\,|G - G_n|^2\,\mathrm{d}s / c^2.$$

If we now choose a zero-approaching sequence $\{c_k\}$ of values of c and a subsequence $\{n_k\}$ of the sequence of natural numbers such that

$$\sum_{k} \int_{t_0}^{T} E\,|G - G_{n_k}|^2\,\mathrm{d}s / c_k^2 < \infty,$$

then

$$\sum_{k} P\,[\sup_{t_0 \le t \le T} |X_t - X_t^{(n_k)}| > c_k] < \infty.$$

Therefore, by the Borel-Cantelli lemma (see section 1.6), there exists, for almost all $\omega \in \Omega$ a $k_0 = k_0\,(\omega)$ such that

$$\sup_{t_0 \le t \le T} |X_t\,(\omega) - X_t^{(n_k)}\,(\omega)| \le c_k \quad \text{for all} \quad k \ge k_0\,(\omega).$$

This makes X_t, with probability 1, the uniform limit of a sequence of continuous functions, and hence continuous itself.

Step 3.   Finally, for arbitrary $G \in M_2\,[t_0,\,T]$, we approximate G with a function G_N, for $N > 0$, defined by

$$G_N\,(t) = \begin{cases} G\,(t) & \text{if } \int_{t_0}^{t} |G|^2\,\mathrm{d}s \le N, \\[2ex] 0 & \text{if } \int_{t_0}^{t} |G|^2\,\mathrm{d}s > N. \end{cases}$$

The process

$$X_t^{(N)} = \int_{t_0}^{t} G_N \, dW$$

is, by virtue of step 2, continuous, and we have

$$X_t(\omega) = X_t^{(N)}(\omega) \text{ for all } t \text{ in } [t_0, T]$$

for all ω in the set

$$A_N = \left\{ \omega : \int_{t_0}^{T} |G|^2 \, ds \leqq N \right\} \subset \Omega.$$

Now, $P(\Omega - A_N)$ can be made arbitrarily small by making N sufficiently large. Consequently, those ω at which $X.(\omega)$ is discontinuous in $[t_0, T]$ have probability 0.

d) For $k = 1$, we have the more precise result

$$E |X_t - X_a|^2 = \int_{a}^{t} E |G(s)|^2 \, ds.$$

At $k = 2$, we prove (5.1.6) again, first for step functions and then for general G, by choosing a sequence of step functions G_n with the property

$$\int_{a}^{t} E |G(s) - G_n(s)|^4 \, ds \to 0$$

(see Gikhman and Skorokhod [5], pp. 385–386). For proof for general k, we refer the reader to Gikhman and Skorokhod [36], pp. 26–27. ∎

5.2 Examples and Remarks

We showed in the example discussed in section 4.2 that only Itô's stochastic integral has the martingale property, from which the simple but nonetheless very useful inequalities (5.1.4) and (5.1.5) follow.

Another consequence of the martingale property is

(5.2.1) **Corollary.** If

$$\int_{t_0}^{T} E |G(s)|^2 \, ds < \infty \, ,$$

the process

$$X_t = \int_{t_0}^{t} G \, dW, \quad t_0 \leqq t \leqq T,$$

has orthogonal increments; that is, for $t_0 \leqq r \leqq s \leqq t \leqq u \leqq T$,

(5.2.2) $\qquad E\,(X_u - X_t)\,(X_s - X_r)' = 0.$

Proof. We either use the martingale property or verify (5.2.2) directly by carrying out the multiplication and applying formula (5.1.3) for the covariance matrix

$$E\,(X_u - X_t)\,(X_s - X_r)' = E\,X_u\,X_s' - E\,X_t\,X_s' - E\,X_u\,X_r' + E\,X_t\,X_r'$$

$$= \left(\int_{t_0}^{s} - \int_{t_0}^{s} - \int_{t_0}^{r} - \int_{t_0}^{r} \right) E\,G\,(v)\,G\,(v)'\,dv$$

$$= 0. \quad \blacksquare$$

It would have been possible to obtain the orthogonality of the increments more quickly from the following general formula:

(5.2.3) **Corollary.** Suppose that G and H belong to $M_2^{d,\,m}\,[t_0, T]$ and that

$$\int_{t_0}^{T} E\,|G|^2\,ds < \infty, \quad \int_{t_0}^{T} E\,|H|^2\,ds < \infty.$$

Then, for any two Borel sets $A, B \subset [t_0, T]$, we have

$$E\left(\int_A G\,dW \right)\left(\int_B H\,dW \right)' = \int_{A \cap B} E\,G\,(u)\,H\,(u)'\,du$$

and, in particular,

$$E\left(\int_{t_0}^{t} G\,dW \right)\left(\int_{t_0}^{s} H\,dW \right)' = \int_{t_0}^{\min\,(t,\,s)} E\,G\,(u)\,H\,(u)'\,du, \quad s, t \in [t_0, T].$$

Proof. From the definition in section 5.1, we have

$$\int_A G\,dW = \int_{t_0}^{T} G\,I_A\,dW, \quad \int_B H\,dW = \int_{t_0}^{T} H\,I_A\,dW.$$

Let us look at the individual matrix elements. We have

$$E\left(\int_{t_0}^{T} I_A\,G_{ij}\,dW^j \right)\left(\int_{t_0}^{T} I_B\,H_{kp}\,dW^p \right) = \delta_{jp} \int_{A \cap B} E\,(G_{ij}\,H_{kp})\,ds$$

(certainly for step functions and hence in general), from which the assertion follows. ∎

(5.2.4) **Remark.** Although, under the assumption of Corollary (5.2.1), the stochastic integral has orthogonal and hence uncorrelated increments, these increments are not in general independent. The case in which $G \in M_2 [t_0, T]$ is independent of ω constitutes an exception. Then,

$$(5.2.5) \qquad X_t = \int_{t_0}^{t} G \, dW,$$

where $G \in M_2 [t_0, T]$ is independent of ω, is, according to Corollary (4.5.6), a *Gaussian process* on the interval $[t_0, T]$ with expectation $E\,X_t = 0$ and with covariance matrix

$$E\,X_t\,X_s' = \int_{t_0}^{\min(t,\,s)} G(u)\,G(u)'\,du.$$

Since uncorrelatedness of normally distributed random variables implies their independence, X_t is a process *with independent increments*.

Conversely, *all* (smooth) d-dimensional Gaussian processes for which $E\,X_t = 0$ and $X_{t_0} = 0$ that have independent increments in the interval $[t_0, T]$ can be represented as stochastic integrals of the form (5.2.5). Specifically, if the variance

$$E\,X_t\,X_t' = Q(t)$$

is an absolutely continuous (or even a continuously differentiable) function of t, we can write

$$E\,X_t\,X_t' = \int_{t_0}^{t} q(s)\,ds, \quad q(s) = \dot{Q}(s).$$

Now, $Q(t)$ is a nonnegative-definite symmetric $d \times d$ matrix that increases monotonically with t; that is,

$$Q(t) - Q(s) \geqq 0, \quad t \geqq s.$$

Therefore, $q(s)$ is a nonnegative-definite symmetric $d \times d$ matrix and thus can be written in the form

$$q(s) = G(s)\,G(s)',$$

where $G(s)$ is a $d \times d$ matrix. For example, we can always take

$$G(s) = U(s)\,\Lambda(s)^{1/2}\,U(s)'$$

or

$$G(s) = U(s)\,\Lambda(s)^{1/2} \ .$$

Here, the $\Lambda\,(s)$ and $U\,(s)$ satisfy

$$q\,(s) = U\,(s)\,\Lambda\,(s)\,U\,(s)' \,;$$

i.e. $\Lambda\,(s)$ is the diagonal matrix of the eigenvalues (arranged in increasing order) and $U\,(s)$ is the orthogonal matrix of (column) eigenvectors of $q\,(s)$. With the first choice, $G\,(s)$ is again nonnegative-definite. Now if W_t is a d-dimensional Wiener process, then the process

$$Y_t = \int_{t_0}^{t} G\,(s)\,\mathrm{d}W_s, \quad t_0 \leqq t \leqq T$$

coincides, with respect to distribution, with the given process X_t, as one can immediately verify by calculating the first two moments.

(5.2.6) Remark. The Gaussian process mentioned in Remark (5.2.4) is, in the case $d = m = 1$ and $t_0 = 0$, "essentially" (that is, up to within a transformation of the time axis) a piece of the Wiener process. To see this, let us set

$$\tau\,(t) = \int_{0}^{t} G\,(s)^2\,\mathrm{d}s, \quad 0 \leqq t \leqq T.$$

Then, there exists a Wiener process $\overline{W}_t$ such that

$$X_t = \int_{0}^{t} G\,\mathrm{d}W = \overline{W}_{\tau\,(t)}$$

or

$$(5.2.7) \qquad X_{\tau^{-1}(t)} = \overline{W}_t,$$

as can be checked immediately by calculating the first two moments. For this reason, we call $\tau\,(t)$ the **intrinsic time** of X_t. Here, $\tau^{-1}\,(t) = \min\,(s : \tau\,(s) = t)$ is defined for $t \leqq \tau\,(T)$.

This consideration can be carried over to the case of arbitrary $G \in M_2^{1,\,m}\,[t_0, T]$ if we define the intrinsic time by

$$\tau\,(t) = \int_{t_0}^{t} |G\,(s)|^2\,\mathrm{d}s, \quad t_0 \leqq t \leqq T.$$

Now, $\tau\,(t)$ is itself a (nonanticipating) random function (see McKean [45], pp. 29–31). From the representation (5.2.7), for example, we immediately derive the law of the iterated logarithm for a one-dimensional stochastic integral

$$X_t = \int_{t_0}^{t} G\,\mathrm{d}W, \quad G \in M_2^{1,\,m}\,[t_0, T],$$

in the form

$$\lim_{t \downarrow t_0} \sup \frac{X_t}{\sqrt{2\,\tau\,(t)\,\log\,\log\,1/\tau\,(t)}} = 1.$$

(5.2.8) **Remark.** The property of having infinite length in every finite interval also carries over from sample functions of W_t to sample functions of the stochastic integral X_t. This follows, as for W_t in section 3.1, from the following more precise result (see Goldstein [37a], Theorem 4.1): Suppose that $G \in M_2^{d,\,m}\,[t_0, T]$, $t_0 < t_1 < \ldots < t_n = T < \infty$, $\delta_n = \max_k\,(t_k - $. Then for

$$X_t = \int_{t_0}^t G \, \mathrm{d}W \, ,$$

we have

(5.2.9) $\mathrm{st}\text{-}\lim\limits_{\delta_n \to 0} \sum\limits_{k=1}^n (X_{t_k} - X_{t_{k-1}})\,(X_{t_k} - X_{t_{k-1}})' = \int\limits_{t_0}^T G\,(s)\,G\,(s)'\,\mathrm{d}s$

and, in particular,

(5.2.10) $\mathrm{st}\text{-}\lim\limits_{\delta_n \to 0} \sum\limits_{k=1}^n |X_{t_k} - X_{t_{k-1}}|^2 = \int\limits_{t_0}^T |G\,(s)|^2\,\mathrm{d}s \, .$

One should note that random variables appear in general in the right-hand members of (5.2.9) and (5.2.10). Thus, we have the alternative: Either $X_t \equiv 0$ (for all $t \in [t_0, T]$), as will be the case if the right-hand member of (5.2.10) vanishes, or X_t is not of bounded variation on $[t_0, T]$, as will be the case if the right-hand member of (5.2.10) fails to vanish.

5.3 Stochastic Differentials. Itô's Theorem.

The relationship

$$X_t\,(\omega) = \int_{t_0}^t G\,(s,\,\omega)\,\mathrm{d}W_s\,(\omega)$$

can also be written as

$$\mathrm{d}X_t = G\,(t)\,\mathrm{d}W_t \, .$$

This is a special so-called **stochastic differential**. We shall now define and investigate such differentials. To do this, let us look at a somewhat more general stochastic process of the form

(5.3.1) $\qquad X_t(\omega) = X_{t_0}(\omega) + \int\limits_{t_0}^{t} f(s, \omega)\, ds + \int\limits_{t_0}^{t} G(s, \omega)\, dW_s(\omega).$

Here, we again assume the usual situation: W_t is an m-dimensional Wiener process, $\mathfrak{F}_t$ is the accompanying family of sigma-algebras with events independent of $\mathfrak{W}_t^+$, and G is a $(d \times m$ matrix)-valued function in $M_2^{d,m}[t_0, T] = M_2[t_0, T]$. Then, the stochastic integral in (5.3.1) is completely defined for $t_0 \leq t \leq T$.

As regards X_{t_0} and f, we now make the following assumptions:

a) X_{t_0} is an $\mathfrak{F}_{t_0}$-measurable random variable (and as such independent of $\mathfrak{W}_t^+$, and hence of $W_t - W_{t_0}$ for $t \geq t_0$). In particular, this is the case when X_{t_0} is not random.

b) The function f is an R^d-valued function measurable in (s, ω) and nonanticipating: that is, $f(t, \cdot)$ is $\mathfrak{F}_t$-measurable for all $t \in [t_0, T]$, and we have with probability 1

$$\int\limits_{t_0}^{T} |f(s, \omega)|\, ds < \infty.$$

We interpret the last integral, like the corresponding integral of f in (5.3.1), as the usual (Lebesgue or possibly Riemann) integral of the sample functions $f(\cdot, \omega)$.

(5.3.2) **Remark.** Both integrals in (5.3.1) are continuous functions of the upper limit (the integral of f is in fact absolutely continuous!), so that the process X_t is an R^d-valued process that, with probability 1, has continuous sample functions. Furthermore, X_t is $\mathfrak{F}_t$-measurable and hence nonanticipating, and we have, for every s such that $t_0 \leq s \leq t \leq T$,

$$X_t = X_s + \int\limits_{s}^{t} f(u)\, du + \int\limits_{s}^{t} G(u)\, dW_u.$$

Stochastic differentials are simply a more compact symbolic notation for relationships of the form (5.3.1).

(5.3.3) **Definition.** We shall say that a stochastic process X_t defined by equation (5.3.1) possesses the **stochastic differential** $f(t)\, dt + G(t)\, dW_t$ and we shall write

(5.3.4) $\qquad dX_t = f(t)\, dt + G(t)\, dW_t$

$\qquad\qquad = f\, dt + G\, dW.$

(5.3.5) **Remark.** In the shift from (5.3.1) to the stochastic differential (5.3.4), the initial value X_{t_0} disappears, so that we can get from (5.3.4) only the differences

$$X_t - X_s = \int\limits_s^t f(u)\,du + \int\limits_s^t G(u)\,dW_u$$

and must always specify X_{t_0} when it is nonzero.

(5.3.6) **Example.** Suppose that $d = m = 1$, that $t_0 = 0$, and that T is an arbitrary positive number. The differential notation of

$$\int\limits_0^t W_s\,dW_s = W_t^2/2 - t/2$$

is

(5.3.7) $d(W_t^2) = dt + 2\,W_t\,dW_t.$

If we construct the differential of W_t^2 formally, using Taylor's theorem, we obtain

$$d(W_t^2) = 2\,W_t\,dW_t + (dW_t)^2.$$

Comparison with (5.3.7) shows that, in the case of the *stochastic* differential of W_t^2, we must regard the *first two* terms as first-order terms and must replace $(dW_t)^2$ with dt (see remark (3.1.9)).

The overall explanation of the phenomenon is to be found in the following theorem of Itô [42]. It says, in the language of stochastic differentials, that smooth functions of processes defined by (5.3.1) are themselves processes of this type. Here is Itô's theorem in its most general form:

(5.3.8) **Itô's Theorem.** Let $u = u(t, x)$ denote a continuous function defined on $[t_0, T] \times R^d$ with values in R^k and with the continuous partial derivatives (k-vectors!)

$$\frac{\partial}{\partial t} u(t, x) = u_t,$$

$$\frac{\partial}{\partial x_i} u(t, x) = u_{x_i}, \quad x = (x_1, \dots, x_d)',$$

$$\frac{\partial^2}{\partial x_i\,\partial x_j} u(t, x) = u_{x_i x_j}, \quad i, j \le d.$$

If the d-dimensional stochastic process X_t is defined on $[t_0, T]$ by the stochastic differential

$$dX_t = f(t)\,dt + G(t)\,dW_t, \quad W_t\ m\text{- dimensional,}$$

then the k-dimensional process

$$Y_t = u(t, X_t)$$

defined on $[t_0, T]$ with initial value $Y_{t_0} = u(0, X_{t_0})$ also possesses a stochastic differential with respect to the *same Wiener process* W_t, and we have

$$(5.3.9a) \qquad dY_t = \left(u_t(t, X_t) + u_x(t, X_t) f(t) + \frac{1}{2} \sum_{i=1}^{d} \sum_{j=1}^{d} u_{x_i x_j}(t, X_t)(G(t) \right.$$
$$\left. G(t)')_{ij} \right) dt + u_x(t, X_t) G(t) dW_t.$$

Here, $u_x = (u_{x_1}, \dots, u_{x_d})$ is a $k \times d$ matrix and $u_{x_i x_j}$ is a k-dimensional column vector.

(5.3.10) **Remark.** The double summation in (5.3.9a) can also be written as follows:

$$\sum_{i=1}^{d} \sum_{j=1}^{d} u_{x_i x_j} (G G')_{ij} = \text{tr}(u_{xx} G G') = \text{tr}(G G' u_{xx}),$$

where $u_{xx} = (u_{x_i x_j})$ is a $d \times d$ matrix whose elements are k-vectors. Then, (5.3.9a) takes the form

$$(5.3.9b) \qquad dY_t = u_t\, dt + u_x\, dX_t + \frac{1}{2} \text{tr}(G G' u_{xx})\, dt.$$

Let us now specialize Theorem (5.3.8) to the case $k = m = 1$, which is important for many applications:

(5.3.11) **Itô's Theorem for $k = m = 1$.** Let $u = u(t, x_1, \dots, x_d)$ denote a continuous function defined on $[t_0, T] \times R^d$ with continuous partial derivatives u_t, u_{x_i} and $u_{x_i x_j}$ for $i, j \leq d$. Furthermore, suppose that d one-dimensional stochastic processes $X_i(t)$ are defined on $[t_0, T]$ by the stochastic differentials

$$dX_i(t) = f_i(t)\, dt + G_i(t)\, dW_t, \quad i = 1, 2, \dots, d,$$

with respect to the same one-dimensional Wiener process. Then, the process

$$Y_t = u(t, X_1(t), \dots, X_d(t))$$

also possesses a stochastic differential on $[t_0, T]$; and

$$dY_t = u_t\, dt + \sum_{i=1}^{d} u_{x_i}\, dX_i + \frac{1}{2} \sum_{i=1}^{d} \sum_{j=1}^{d} u_{x_i x_j}\, dX_i\, dX_j.$$

Here, the product $dX_i\, dX_j$ can be calculated from the following multiplication table:

$\times$	dW	dt
dW	dt	0
dt	0	0

In other words,

$$dX_i \, dX_j = G_i \, G_j \, dt, \quad i, j \leqq d,$$

and

$$dY_t = \left(u_t + \sum_{i=1}^{d} u_{x_i} f_i + \frac{1}{2} \sum_{i=1}^{d} \sum_{j=1}^{d} u_{x_i x_j} \, G_i \, G_j\right) dt + \left(\sum_{i=1}^{d} u_{x_i} \, G_i\right) dW_t.$$

We single out the even more special case $k = m = d = 1$ as

(5.3.12) **Corollary** (Itô's theorem for $k = m = d = 1$). Let $u = u\,(t, x)$ denote a scalar continuous function defined on $[t_0, T] \times R^1$ with continuous partial derivatives u_t, u_x, and u_{xx}. If X_t is a process defined on $[t_0, T]$ with stochastic differential

$$dX_t = f \, dt + G \, dW,$$

where f, G, and W_t are scalar functions, then $Y_t = u\,(t, X_t)$ possesses on $[t_0, T]$ the stochastic differential

$$dY_t = \left(u_t(t, X_t) + u_x\,(t, X_t)\, f\,(t) + \frac{1}{2}\, u_{xx}\,(t, X_t)\, G\,(t)^2\right) dt$$

$$+ u_x\,(t, X_t)\, G\,(t)\, dW_t.$$

Before we give a proof of Theorem (5.3.8), let us make clear its scope and usefulness with various examples.

5.4 Examples and Remarks in Connection with Itô's Theorem

The noteworthy, as compared with the usual differential extra term in (5.3.9) is formed from the second derivatives $u_{x_i x_j}$. This term is the most frequent cause of errors in purely formal manipulation of stochastic differential equations.

(5.4.1) **Example.** Theorem (5.3.11) yields in the case

$$u = x_1 \, x_2$$

the following result: If

$$dX_1\,(t) = f_1\,(t)\, dt + G_1\,(t)\, dW_t,$$
$$dX_2\,(t) = f_2\,(t)\, dt + G_2\,(t)\, dW_t,$$

then

$$d\,(X_1\,(t)\, X_2\,(t)) = X_1\,(t)\, dX_2\,(t) + X_2\,(t)\, dX_1\,(t) + G_1\,(t)\, G_2\,(t)\, dt$$
$$= (X_1\, f_2 + X_2\, f_1 + G_1\, G_2)\, dt + (X_1\, G_2 + X_2\, G_1)\, dW_t.$$

This is the rule for integration of stochastic integrals by parts. In integral form, it is

$$X_1(t) X_2(t) = X_1(t_0) X_2(t_0) + \int_{t_0}^{t} X_1 \, dX_2 + \int_{t_0}^{t} X_2 \, dX_1 + \int_{t_0}^{t} G_1 G_2 \, ds,$$

$$t_0 \leq t \leq T.$$

In comparison with the corresponding formulas for ordinary integrals or differentials, there is the extra term

$$G_1 G_2 (dW)^2 = G_1 G_2 \, dt.$$

The choice $X_1(t) = t$, $X_2(t) = W_t$ yields

$$d(t W_t) = W_t \, dt + t \, dW_t,$$

and the choice $X_1(t) = X_2(t) = W_t$ yields the familiar result

$$d(W_t^2) = dt + 2 W_t \, dW_t.$$

(5.4.2) **Example.** We shall study, in particular, smooth functions of the Wiener process itself. For the scalar situation, Corollary (5.3.12) yields with $X_t = W_t$ and $0 \leq t < \infty$

$$du(t, W_t) = \left(u_t(t, W_t) + \frac{1}{2} u_{xx}(t, W_t) \right) dt + u_x(t, W_t) \, dW_t.$$

In the special case in which $u = u(x)$ is independent of t and twice continuously differentiable with respect to x, we obtain

(5.4.3a) $$du(W_t) = u'(W_t) \, dW_t + \frac{1}{2} u''(W_t) \, dt$$

or, what amounts to the same thing,

(5.4.3b) $$u(W_t) = u(0) + \int_{0}^{t} u'(W_s) \, dW_s + \frac{1}{2} \int_{0}^{t} u''(W_s) \, ds.$$

The most interesting term in (5.4.3b) is the stochastic integral with respect to W_t, for which we have thus found an expression containing only an ordinary integral.

Formulas (5.4.3a) and (5.4.3b) bring out sharply the essential characteristic of the calculus of stochastic integrals, specifically, the presence of an extra first-order term in the differential of smooth functions of a Wiener process W_t. Equation (5.4.3b) is sometimes called the "fundamental theorem of the calculus of (Ito's) stochastic integrals."

(5.4.4) **Example.** For $u(x) = x^n$, where $n = 1, 2, \ldots$, and $t \geq 0$, formula (5.4.3a) yields

$$d\left(W_t^n\right) = n\,W_t^{n-1}\,dW_t + \frac{n\,(n-1)}{2}\,W_t^{n-2}\,dt\,.$$

(5.4.5) **Example.** Let us look again at the one-dimensional case $d = m = 1$. We begin with the process

$$X_t = X_{t_0} - \frac{1}{2}\int_{t_0}^{t} G\,(s)^2\,ds + \int_{t_0}^{t} G\,(s)\,dW_s\,,\quad G \in M_2^{1,\,1}\,[t_0,\,T]\,,$$

and calculate the stochastic differential for the process

$$Y_t = e^{X_t}\,.$$

For $u\,(x) = e^x$, Corollary (5.3.12) yields

$$dY_t = e^{X_t}\,G\,(t)\,dW_t$$

or

(5.4.6) $dY_t = Y_t\,G\,(t)\,dW_t\,,\quad Y_{t_0} = e^{X_{t_0}} = c > 0\,.$

This is a stochastic differential equation for the process Y_t with initial condition $Y_{t_0} > 0$. From the above-given derivation of the equation, we know that the process

$$Y_t = Y_{t_0}\exp\left(-\frac{1}{2}\int_{t_0}^{t} G\,(s)^2\,ds + \int_{t_0}^{t} G\,(s)\,dW_s\right)$$

satisfies this equation for $t \in [t_0,\,T]$.

For $G \equiv 1$ and $t_0 = 0$, we have the result that the equation

(5.4.7) $dY_t = Y_t\,dW_t\,,\quad Y_0 = 1\,,$

has the solution

$$Y_t = \exp\,(W_t - t/2),\quad t \geqq 0\,.$$

If we interpret equation (5.4.7) as an ordinary differential equation for continuously differentiable functions, we obtain $Y_t = c\,\exp\,W_t$ as its solution. Therefore, we can say that the role of the usual exponential function in the calculus of stochastic differentials is taken over by the function $\exp\,(W_t - t/2)$.

(5.4.8) **Remark.** Let us return to the general equation (5.3.1). If X_{t_0} has a normal distribution or if it is a constant and if the functions f and G are *independent of* ω, we can generalize Remark (5.2.4) as follows: The process

$$X_t\,(\omega) = X_{t_0}\,(\omega) + \int_{t_0}^{t} f\,(s)\,ds + \int_{t_0}^{t} G\,(s)\,dW_s\,(\omega),\quad t_0 \leqq t \leqq T\,,$$

is a d-dimensional Gaussian process with independent increments, with expectation

$$E\,X_t = E\,X_{t_0} + \int_{t_0}^{t} f\,\mathrm{d}s$$

and with covariance matrix

$$E\,(X_t - E\,X_t)\,(X_s - E\,X_s)' = \mathrm{Cov}\,(X_{t_0}, X'_{t_0}) + \int_{t_0}^{\min\,(t,\,s)} G\,(u)\,G\,(u)'\,\mathrm{d}u.$$

Conversely, all smooth d-dimensional Gaussian processes with independent increments can be represented in this form. Specifically, if

$$E\,(X_t - X_{t_0}) = F\,(t), \quad t_0 \leqq t \leqq T,$$

is, for example, continuously differentiable, we can write

$$E\,(X_t - X_{t_0}) = \int_{t_0}^{t} f\,(s)\,\mathrm{d}s, \quad f\,(s) = \dot{F}\,(s).$$

The process

$$Y_t = X_t - X_{t_0} - \int_{t_0}^{t} f\,(s)\,\mathrm{d}s$$

satisfies the assumptions $Y_{t_0} = 0$ and $E\,Y_t = 0$ made in remark (5.2.4). Therefore, if the $d \times d$ matrix

$$E\,Y_t\,Y'_t = Q\,(t)$$

is, for example, continuously differentiable, there exists at least one $d \times d$ matrix G such that

$$Q\,(t) = \int_{t_0}^{t} q\,(s)\,\mathrm{d}s = \int_{t_0}^{t} G\,(s)\,G\,(s)'\,\mathrm{d}s, \quad q\,(s) = \dot{Q}\,(s).$$

Then, if W_t is a d-dimensional Wiener process such that X_{t_0} and $W_t - W_{t_0}$ are, for $t \geqq t_0$, independent, the process

$$Z_t = X_{t_0} + \int_{t_0}^{t} f\,(s)\,\mathrm{d}s + \int_{t_0}^{t} G\,(s)\,\mathrm{d}W_s, \quad t_0 \leqq t \leqq T,$$

coincides distributionwise with X_t. It should be noted that W_t can in general be m-dimensional (where $m \gtreqqless d$). It is only necessary to choose a $d \times m$ matrix G such that $\dot{Q}\,(t) = G\,(t)\,G\,(t)'$.

Some authors (see, for example, Bucy and Joseph [61]) have analyzed stochastic differentials in terms of an m-dimensional Gaussian process V_t with independent increments $V_0 = 0$, $E V_t = 0$, and covariance

$$E V_t V_s' = \int_0^{\min(t, s)} q(u) \, du.$$

However, by virtue of what was said above, we can always represent V_t in the form

$$dV_t = G_0(t) \, dW_t, \quad G_0(t) G_0(t)' = q(t).$$

Therefore,

$$dX_t = f \, dt + G \, dV_t = f \, dt + G G_0 \, dW_t;$$

that is, we can confine ourselves to differentials with respect to W_t.

(5.4.9)   **Example.** $u(x) = |x|^2 = x' x$ yields with (5.3.9b) for $k = 1$ and arbitrary d and m

$$d|X_t|^2 = 2 X_t' \, dX_t + |G(t)|^2 \, dt.$$

5.5 Proof of Itô's Theorem

Since the proof of Theorem (5.3.8) differs from that of Corollary (5.3.12) only in the more complicated notation, we shall confine ourselves to the proof of the corollary. We can in fact simplify the situation still further without changing the basic idea of the proof.

Itô's differential formula is a short notation for an integral expression for the process $Y_t = u(t, X_t)$. It will be sufficient to prove the formula for step functions f and G. As usual, the general case can be obtained by a passage to the limit. Since the domain of definition of a step function is decomposed into finitely many intervals in each of which it is constant (as a function of t), we can confine ourselves to the case of constant $f(t, \omega) \equiv f(\omega)$ and $G(t, \omega) \equiv G(\omega)$.

Thus, our initial process X_t has the form

$$X_t = X_{t_0} + f(t - t_0) + G(W_t - W_{t_0}), \quad t_0 \le t \le T,$$

where X_{t_0}, f, and G are random variables. The $\mathfrak{F}_{t_0}$-measurability of X_{t_0}, f, and G implies that they are *independent* of $W_t - W_{t_0}$ for $t \ge t_0$ (though otherwise arbitrary).

The process Y_t has the form

$$Y_t = u(t, X_t) = u(t, X_{t_0} + f(t - t_0) + G(W_t - W_{t_0}))$$

with

$$Y_{t_0} = u(t_0, X_{t_0}).$$

Suppose that $t_0 < t_1 < \ldots < t_n = t \leqq T$. Then,

(5.5.1) $\qquad Y_t - Y_{t_0} = \sum_{k=1}^{n} (u(t_k, X_{t_k}) - u(t_{k-1}, X_{t_{k-1}})).$

With our assumptions on $u(t, x)$, Taylor's formula yields

$$u(t_k, X_{t_k}) - u(t_{k-1}, X_{t_{k-1}}) = u_t(t_{k-1} + d_k(t_k - t_{k-1}), X_{t_{k-1}})(t_k - t_{k-1})$$

(5.5.2) $\qquad + u_x(t_{k-1}, X_{t_{k-1}})(X_{t_k} - X_{t_{k-1}})$

$$+ \frac{1}{2} u_{xx}(t_{k-1}, X_{t_{k-1}} + \bar{d}_k(X_{t_k} - X_{t_{k-1}}))(X_{t_k} - X_{t_{k-1}})^2,$$

where $0 < d_k$ and $\bar{d}_k < 1$. In view of the continuity of X_t, u_t, and u_{xx}, we see that there exist random variables α_n and β_n that converge with probability 1 to 0 as

$$\delta_n = \max_{1 \leqq k \leqq n} (t_k - t_{k-1}) \longrightarrow 0$$

and that satisfy the inequalities

$$\max_{1 \leqq k \leqq n} |u_t(t_{k-1} + d_k(t_k - t_{k-1}), X_{t_{k-1}}) - u_t(t_{k-1}, X_{t_{k-1}})| \leqq \alpha_n$$

and

$$\max_{1 \leqq k \leqq n} |u_{xx}(t_{k-1}, X_{t_{k-1}} + \bar{d}_k(X_{t_k} - X_{t_{k-1}})) - u_{xx}(t_{k-1}, X_{t_{k-1}})| \leqq \beta_n.$$

Consequently, if we substitute (5.5.2) into (5.5.1) and replace d_k and $\bar{d}_k$ with 0, since

$$\sum_{k=1}^{n} (t_k - t_{k-1}) = t - t_0$$

and

$$\text{st-}\lim_{\delta_n \to 0} \sum_{k=1}^{n} (X_{t_k} - X_{t_{k-1}})^2 = G^2(t - t_0),$$

this does not change the limit of $Y_t - Y_{t_0}$ in (5.5.1) as $\delta_n \to 0$. What we need to show, then, is that

$$\text{st-}\lim_{\delta_n \to 0} \sum_{k=1}^{n} \Big[u_t(t_{k-1}, X_{t_{k-1}})(t_k - t_{k-1}) + u_x(t_{k-1}, X_{t_{k-1}})(X_{t_k} - X_{t_{k-1}})$$

$$+ \frac{1}{2} u_{xx}(t_{k-1}, X_{t_{k-1}})(X_{t_k} - X_{t_{k-1}})^2 \Big] =$$

$$= \int_{t_0}^{t} \left(u_t(s, X_s) + u_x(s, X_s) f + \frac{1}{2} u_{xx}(s, X_s) G^2 \right) ds + \int_{t_0}^{t} u_x(s, X_s) G \, dW_s.$$

By virtue of the continuity assumptions, we have

$$\underset{\delta_n \to 0}{\text{ac-lim}} \sum_{k=1}^{n} u_t(t_{k-1}, X_{t_{k-1}})(t_k - t_{k-1}) = \int_{t_0}^{t} u_t(s, X_s) \, ds$$

and

$$\underset{\delta_n \to 0}{\text{st-lim}} \sum_{k=1}^{n} u_x(t_{k-1}, X_{t_{k-1}})(X_{t_k} - X_{t_{k-1}}) = \int_{t_0}^{t} u_x(s, X_s) f \, ds +$$

$$\int_{t_0}^{t} u_x(s, X_s) G \, dW_s.$$

We still need to take care of the sum

$$\sum_{k=1}^{n} u_{xx}(t_{k-1}, X_{t_{k-1}})(X_{t_k} - X_{t_{k-1}})^2 = f^2 \sum_{k=1}^{n} u_{xx}(t_{k-1}, X_{t_{k-1}})(t_k - t_{k-1})^2$$

$$+ 2 f G \sum_{k=1}^{n} u_{xx}(t_{k-1}, X_{t_{k-1}})(t_k - t_{k-1})(W_{t_k} - W_{t_{k-1}})$$

$$+ G^2 \sum_{k=1}^{n} u_{xx}(t_{k-1}, X_{t_{k-1}})(W_{t_k} - W_{t_{k-1}})^2.$$

Since the first two sums in the right-hand member converge, by virtue of the continuity of u_{xx} and W_t, to 0 with probability 1, it remains only to show that

$$(5.5.3) \qquad \underset{\delta_n \to 0}{\text{st-lim}} \sum_{k=1}^{n} u_{xx}(t_{k-1}, X_{t_{k-1}})(W_{t_k} - W_{t_{k-1}})^2 = \int_{t_0}^{t} u_{xx}(s, X_s) \, ds.$$

Since

$$\underset{\delta_n \to 0}{\text{ac-lim}} \sum_{k=1}^{n} u_{xx}(t_{k-1}, X_{t_{k-1}})(t_k - t_{k-1}) = \int_{t_0}^{t} u_{xx}(s, X_s) \, ds \, ,$$

(5.5.3) reduces to

$$\underset{\delta_n \to 0}{\text{st-lim}} \, S_n = 0$$

with

$$S_n = \sum_{k=1}^{n} u_{xx}(t_{k-1}, X_{t_{k-1}})((W_{t_k} - W_{t_{k-1}})^2 - (t_k - t_{k-1})).$$

We now eliminate large values of u_{xx} by a truncation technique. For positive N, let us define

$$I_k^N(\omega) = \begin{cases} 1 & \text{if } |X_{t_i}| \leqq N \text{ for all } i \leqq k, \\ 0 & \text{otherwise,} \end{cases}$$

$$\varepsilon_k = (W_{t_k} - W_{t_{k-1}})^2 - (t_k - t_{k-1})$$

and

$$S_n^N = \sum_{k=1}^n u_{xx}(t_{k-1}, X_{t_{k-1}}) I_{k-1}^N \varepsilon_k.$$

Since $E\, \varepsilon_k = 0$ and $E\, \varepsilon_k^2 = 2\,(t_k - t_{k-1})^2$ and the ε_k are independent of each other and of $u_{xx}(t_{k-1}, X_{t_{k-1}}) I_{k-1}^N$, we have

$$E\, S_n^N = 0$$

and

$$E\,(S_n^N)^2 = \sum_{k=1}^n E\,(u_{xx}(t_{k-1}, X_{t_{k-1}}) I_{k-1}^N)^2\, E\,\varepsilon_k^2$$

$$\leqq 2 \max_{t_0 \leqq s \leqq t, |y| \leqq N} |u_{xx}(s,y)| \sum_{k=1}^n (t_k - t_{k-1})^2$$

$$\to 0 \quad (\delta_n \to 0).$$

For every fixed $N > 0$, we therefore have

$$\operatorname*{qm\text{-}lim}_{\delta_n \to 0} S_n^N = \operatorname*{st\text{-}lim}_{\delta_n \to 0} S_n^N = 0.$$

The error resulting from the truncation is

(5.5.4) $$P\,[S_n \neq S_n^N] = P\,[\max_{t_0 \leqq s \leqq t} |X_s| > N].$$

Now, the quantity

$$\max_{t_0 \leqq s \leqq t} |X_s| = \max_{t_0 \leqq s \leqq t} |X_{t_0} + f(s - t_0) + G\,(W_s - W_{t_0})|$$

$$\leqq |X_{t_0}| + |f|\,(t - t_0) + |G| \max_{t_0 \leqq s \leqq t} |W_s - W_{t_0}|$$

is an almost certainly finite random variable, so that the right-hand member of (5.5.4) can be made arbitrarily small by choosing N sufficiently great. Since

$$P\,[|S_n| > \varepsilon] \leqq P\,[|S_n^N| > \varepsilon] + P\,[S_n \neq S_n^N]$$

we have

$$\operatorname*{st\text{-}lim}_{\delta_n \to 0} S_n = 0. \; \blacksquare$$

Chapter 6

Stochastic Differential Equations, Existence and Uniqueness of Solutions

6.1 Definition and Examples

Let us look at a stochastic differential of the form

(6.1.1a) $dX_t = f(t, X_t)\, dt + G(t, X_t)\, dW_t, \quad X_{t_0} = c, \quad t_0 \leqq t \leqq T < \infty,$

or, in integral form,

(6.1.1b) $X_t = c + \displaystyle\int_{t_0}^{t} f(s, X_s)\, ds + \int_{t_0}^{t} G(s, X_s)\, dW_s, \quad t_0 \leqq t \leqq T < \infty,$

where X_t is an R^d-valued stochastic process (for the time being, assumed known) defined on $[t_0, T]$ and W_t is an m-dimensional Wiener process. The R^d-valued function f and the $(d \times m$ matrix$)$-valued function G are assumed to be defined and measurable on $[t_0, T] \times R^d$. For fixed (t, x), suppose that $f(t, x)$ and $G(t, x)$ are independent of $\omega \in \Omega$, i.e., that the random parameter ω appears only indirectly in the coefficients in equation (6.1.1) in the form $f(t, X_t(\omega))$ and $G(t, X_t(\omega))$. For a generalization, see remark (6.1.5).

Here, the process X_t, which we assume known, must of course be constructed in such a way that, after it is substituted into (6.1.1a), the right-hand member will become a stochastic differential in the sense of section 5.3. In particular, X_t must not anticipate (see remark (5.3.2)); that is, it must be $\mathfrak{F}_t$-measurable.

Equations (6.1.1) can also be interpreted as the defining equations for an unknown stochastic process X_t with given initial value $X_{t_0} = c$. With regard to the accompanying family of sigma-algebras $\mathfrak{F}_t$, we make once and for all the following

(6.1.2) **Convention.** For the purpose of treating stochastic differential equations on the interval $[t_0, T]$, it is always sufficient to choose for $\mathfrak{F}_t$ the smallest sigma-algebra with respect to which the initial value c and the random variables W_s, for $s \leqq t$, are measurable, specifically,

$$\mathfrak{F}_t = \mathfrak{A}(c; W_s, s \leqq t).$$

By definition, $\mathfrak{F}_t$ must, for all $t \geq t_0$, be independent of

$$\mathfrak{W}_t^+ = \mathfrak{A}\,(W_s - W_t,\, s \geq t)$$

For $t = t_0$, this means in particular that *the initial value c and the Wiener process* $W_t - W_{t_0}$ *must be statistically independent.*

In particular, if c is with probability 1 constant, the independence of $W_t - W_{t_0}$ is trivially satisfied. In this case, we have (except for events of probability 0)

$$\mathfrak{F}_t = \mathfrak{W}\,[0,\,t] = \mathfrak{A}\,(W_s,\, s \leq t).$$

Of course, $\mathfrak{F}_t$ can always be augmented with all events that are independent of $\mathfrak{W}_t^+$. When $\mathfrak{F}_t$ is not specifically identified, we shall always assume the above choice to have been made.

(6.1.3) **Definition.** An equation of the form (6.1.1a) is called an **(Itô's) stochastic differential equation.** The random variable c is called the **initial value** at the instant t_0. Here, (6.1.1a), together with the initial value, is only a symbolic way of writing the stochastic integral equation (6.1.1b). A stochastic process X_t is called a **solution** of equation (6.1.1a) or (6.1.1b) on the interval $[t_0,\, T]$ if it has the following properties:

a) X_t is $\mathfrak{F}_t$-measurable, that is, nonanticipating for $t \in [t_0,\, T]$.

b) The functions $\overline{f}\,(t,\,\omega) = f\,(t,\, X_t\,(\omega))$ and $\overline{G}\,(t,\,\omega) = G\,(t,\, X_t\,(\omega))$ (nonanticipating in accordance with a) are such that, with probability 1,

$$\int_{t_0}^{T} |\overline{f}\,(s,\,\omega)|\,\mathrm{d}s < \infty$$

and

$$\int_{t_0}^{T} |\overline{G}\,(s,\,\omega)|^2\,\mathrm{d}s < \infty$$

(that is, $\overline{G}$ belongs to $M_2^{d,\,m}\,[t_0,\, T]$). Then, in accordance with section 5.3, the right-hand member of (6.1.1a) is meaningful.

c) Equation (6.1.1b) holds for every $t \in [t_0,\, T]$ with probability 1.

(6.1.4) **Remark.** We therefore have the following situation: There are, on the one hand, the fixed functions f and G that determine the "system" and, on the other hand, the two independent random elements c and $W_.$. For almost every choice of $c\,(\omega)$ and almost every Wiener sample function $W_.\,(\omega)$, we obtain via f and G, in the case of a unique solution of (6.1.1), the sample function $X_.\,(\omega)$ of a new process defined on $[t_0,\, T]$ that satisfies (6.1.1). In accordance with (6.1.3a) and our definition of $\mathfrak{F}_t$, X_t is a functional of c and W_s for $s \leqq t$; that is, there exists a function g (uniquely determined by f and G alone) such that

$$X_t = g\,(c\,;W_s,\,s \leqq t).$$

Thus, (6.1.1) can be interpreted as a formula (in general very complicated) determined by the functions f and G with the aid of which the process $X.$ can be constructed from c and $W..$ Here, only $c\,(\omega)$ and the values of $W_s\,(\omega)$, for $s \leqq t$, are used for construction of the value $X_t\,(\omega)$.

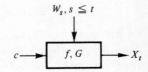

Fig. 6.
X_t as a function of c and W_s, for $s \leqq t$.

(6.1.5) Remark. Stochastic differential equations of the form

$$\mathrm{d}Y_t = f\,(t, Y_t, W_t)\,\mathrm{d}t + G\,(t, Y_t, W_t)\,\mathrm{d}W_t, \quad Y_{t_0} = c,$$

can be converted to the type (6.1.1a) by adding the equation

$$\mathrm{d}W_t = \mathrm{d}W_t$$

and shifting to the $(d+m)$-dimensional state vector

$$X_t = \begin{pmatrix} Y_t \\ W_t \end{pmatrix},$$

and hence to the equation

$$\mathrm{d}X_t = \begin{pmatrix} f \\ 0 \end{pmatrix}\mathrm{d}t + \begin{pmatrix} G \\ I_m \end{pmatrix}\mathrm{d}W_t, \quad X_{t_0} = \begin{pmatrix} c \\ 0 \end{pmatrix}.$$

The coefficients f and G can depend on ω in a more general manner as long as they are nonanticipating. For further details, see Gikhman and Skorokhod [36], pp. 50-53.

(6.1.6) Remark. An nth-order differential equation of the form

$$Y_t^{(n)} = f\,(t, Y_t, \dots, Y_t^{(n-1)}) + G\,(t, Y_t, \dots, Y_t^{(n-1)})\,\xi_t$$

with initial values $Y_{t_0}^{(i)} = c_i$ for $i = 0, 1, \dots, n-1$, where Y_t is R^d-valued and ξ_t is an m-dimensional white noise, can be converted in the usual manner by

$$\mathrm{d}X_t = \mathrm{d}\begin{pmatrix} Y_t \\ \dot{Y}_t \\ \vdots \\ Y_t^{(n-1)} \end{pmatrix} = \begin{pmatrix} \dot{Y}_t \\ \ddot{Y}_t \\ \vdots \\ f\,(t, Y_t, \dots, Y_t^{(n-1)}) \end{pmatrix}\mathrm{d}t + \begin{pmatrix} 0 \\ 0 \\ \vdots \\ G\,(t, Y_t, \dots, Y_t^{(n-1)}) \end{pmatrix}\mathrm{d}W_t$$

into a first-order stochastic differential equation of the type (6.1.1a) for the R^{dn}-valued process X_t with initial value $X_{t_0} = (c_0, \ldots, c_{n-1})'$. We shall consider the case $n = 2$ in detail in section 7.1 (example (7.1.6)).

(6.1.7) **Remark.** Equation (6.1.1b) is equivalent to

$$X_t - X_s = \int\limits_s^t f(u, X_u)\, du + \int\limits_s^t G(u, X_u)\, dW_u, \quad X_{t_0} = c,$$

where $t_0 \leqq s \leqq t \leqq T$. From this it follows that, if $X_t(t_0, c)$ satisfies equation (6.1.1b), then the "semigroup property"

$$X_t(t_0, c) = X_t(s, X_s(t_0, c)), \quad t_0 \leqq s \leqq t \leqq T.$$

holds.

(6.1.8) **Remark.** If X_t is a solution of (6.1.1), then every process stochastically equivalent to X_t is also a solution. Specifically, if, for every fixed $t \in [t_0, T]$,

$$X_t = \overline{X}_t$$

with probability 1 (where the exceptional set must belong to $\mathfrak{F}_t$), then, by virtue of the tacitly assumed separability of all processes, we have, for almost all ω,

$$X_.(\omega) = \overline{X}_.(\omega) \quad \text{in } [t_0, T].$$

This implies

$$\int\limits_{t_0}^t f(s, X_s)\, ds = \int\limits_{t_0}^t f(s, \overline{X}_s)\, ds$$

and, by virtue of Corollary (4.5.4),

$$\int\limits_{t_0}^t G(s, X_s)\, dW_s = \int\limits_{t_0}^t G(s, \overline{X}_s)\, dW_s.$$

Substitution of a solution (which for the moment we assume to exist) into the right-hand member of (6.1.1b) yields, in accordance with remark (5.3.2), a continuous function of t, which, being a solution, is at the same time almost certainly equal to the left-hand member. It follows that, for every solution of (6.1.1), there exists a stochastically equivalent solution with almost certainly continuous sample functions. *Therefore, we shall always consider continuous solutions of stochastic differential equations.*

(6.1.9) **Example.** If $G \equiv 0$, the fluctuational term in (6.1.1) disappears. We interpret (6.1.1) as the ordinary differential equation

$$\dot{X}_t = f(t, X_t), \quad t_0 \leqq t \leqq T,$$

with initial condition $X_{t_0} = c$. A random influence can show up only in the initial value c.

(6.1.10) **Example.** If the functions $f(t, x) \equiv f(t)$ and $G(t, x) \equiv G(t)$ are independent of $x \in R^d$ and if $f \in L_1[t_0, T]$ and $G \in L_2[t_0, T]$, then

$$dX_t = f(t) \, dt + G(t) \, dW_t$$

is a stochastic differential whose coefficients are independent of X_t and hence independent of ω. Therefore, in $[t_0, T]$, the unique solution of (6.1.1) is

$$X_t(\omega) = c(\omega) + \int_{t_0}^t f(s) \, ds + \int_{t_0}^t G(s) \, dW_s(\omega).$$

In accordance with remark (5.4.8), the process X_t is, in the case of normally distributed or constant c, a d-dimensional continuous Gaussian process with independent increments, with expectation

$$E X_t = E c + \int_{t_0}^t f(s) \, ds$$

and with covariance matrix ($X_{t_0} = c$ and $X_t - X_{t_0}$ are independent!)

$$E(X_t - E X_t)(X_s - E X_s)' = \operatorname{Cov}(c, c') + \int_{t_0}^{\min(t, s)} G(u) G(u)' \, du,$$

for $s, t \in [t_0, T]$.

(6.1.11) **Example.** The example (5.4.5) shows that, if we choose $d = m = 1$ and a function $g \in L_2[t_0, T]$, the process

$$X_t = \exp\left(-\frac{1}{2} \int_{t_0}^t g(s)^2 \, ds + \int_{t_0}^t g(s) \, dW_s\right)$$

is a solution of the equation

$$dX_t = g(t) X_t \, dW_t, \quad X_{t_0} = 1.$$

In this case, $f(t, x) \equiv 0$ and $G(t, x) = g(t) x$. The uniqueness of the solution for all functions g that are bounded on $[t_0, T]$ follows from section 6.2. For the special case $g \equiv 1$, the equation

$$dX_t = X_t \, dW_t, \quad X_{t_0} = 1,$$

has, for every interval $[t_0, T] \subset [0, \infty)$, the solution

$$X_t = \exp(W_t - W_{t_0} - (t - t_0)/2), \quad t \geq t_0.$$

6.2 Existence and Uniqueness of a Solution

To ensure existence and uniqueness of a solution of an ordinary differential equation

(6.2.1) $\qquad \dot{X}_t = f(t, X_t), \quad X_{t_0} = c,$

on the interval $[t_0, T]$, we usually assume that $f(t, x)$ satisfies a so-called Lipschitz condition in x and boundedness with respect to t for every x. These conditions ensure that the Picard-Lindelöf iteration procedure

$$X_t^{(n)} = c + \int_{t_0}^t f(s, X_s^{(n-1)}) \, ds, \quad X_t^{(0)} = c,$$

will converge to a solution of the integral equation

$$X_t = c + \int_{t_0}^t f(s, X_s) \, ds$$

which is equivalent to (6.2.1). Since an ordinary differential equation is a special case (with $G \equiv 0$) of a stochastic differential equation and we wish to obtain a solution of a stochastic differential equation by means of a similar iteration procedure, our sufficient conditions are modeled after the classical ones. We have the following theorem, which we shall refer to in what follows as the existence-and-uniqueness theorem.

(6.2.2) **Theorem.** Suppose that we have a stochastic differential equation

(6.2.3) $\qquad dX_t = f(t, X_t) \, dt + G(t, X_t) \, dW_t, \quad X_{t_0} = c, \quad t_0 \leqq t \leqq T < \infty,$

where W_t is an R^m-valued Wiener process and c is a random variable independent of $W_t - W_{t_0}$ for $t \geqq t_0$. Suppose that the R^d-valued function $f(t, x)$ and the $(d \times m$ matrix)-valued function $G(t, x)$ are defined and measurable on $[t_0, T] \times R^d$ and have the following properties: There exists a constant $K > 0$ such that

a) (Lipschitz condition) for all $t \in [t_0, T]$, $x \in R^d$, $y \in R^d$,

(6.2.4) $\qquad |f(t, x) - f(t, y)| + |G(t, x) - G(t, y)| \leqq K |x - y|$

$(|G|^2 = \operatorname{tr} G G').$

b) (Restriction on growth) For all $t \in [t_0, T]$ and $x \in R^d$,

(6.2.5) $\qquad |f(t, x)|^2 + |G(t, x)|^2 \leqq K^2 (1 + |x|^2).$

Then, equation (6.2.3) has on $[t_0, T]$ a unique R^d-valued solution X_t, continuous with probability 1, that satisfies the initial condition $X_{t_0} = c$; that is, if X_t and Y_t are continuous solutions of (6.2.3) with the same initial value c, then

$$P[\sup_{t_0 \leqq t \leqq T} |X_t - Y_t| > 0] = 0.$$

Proof: First we shall prove the uniqueness; then, with the aid of an iteration procedure, we shall prove that a solution exists.

a) Uniqueness. Let X_t and Y_t denote two continuous solutions of (6.2.3). We would like to show that

$$E\,|X_t - Y_t|^2 = 0 \quad \text{for all} \quad t \in [t_0, T].$$

However, since the second moments of X_t and Y_t are not necessarily finite, we must again work with a truncation procedure. Suppose that, for $N > 0$ and $t \in [t_0, T]$,

$$I_N(t) = \begin{cases} 1, & \text{if } |X_s| \leq N \text{ and } |Y_s| \leq N \text{ for } t_0 \leq s \leq t, \\ 0, & \text{otherwise.} \end{cases}$$

Since

$$I_N(t) = I_N(t)\,I_N(s), \quad s \leq t,$$

we have

(6.2.6) $$I_N(t)\,(X_t - Y_t) = I_N(t)\left(\int_{t_0}^{t} I_N(s)\,(f(s, X_s) - f(s, Y_s))\,ds\right.$$

$$\left. + \int_{t_0}^{t} I_N(s)\,(G(s, X_s) - G(s, Y_s))\,dW_s\right).$$

The last integral is meaningful since $I_N(s)$ is nonanticipating (and, by virtue of the assumption on X_s and Y_s, so are $G(s, X_s)$ and $G(s, Y_s)$ [see remark (4.3.8)]).

For $s \in [t_0, t]$, the Lipschitz condition (6.2.4) implies

$$I_N(s)\,(|f(s, X_s) - f(s, Y_s)| + |G(s, X_s) - G(s, Y_s)|)$$

$$\leq K\,I_N(s)\,|X_s - Y_s| \leq 2\,K\,N.$$

Therefore, the second moments of the two integrals in (6.2.6) exist. By virtue of the inequality $|x + y|^2 \leq 2\,(|x|^2 + |y|^2)$, Schwarz's inequality, and formula (5.1.3a), we obtain from (6.2.6)

$$E\,I_N(t)\,|X_t - Y_t|^2 \leq 2\,E \left|\int_{t_0}^{t} I_N(s)\,(f(s, X_s) - f(s, Y_s))\,ds\right|^2$$

$$+ 2\,E\left|\int_{t_0}^{t} I_N(s)\,(G(s, X_s) - G(s, Y_s))\,dW_s\right|^2$$

$$\leq 2\,(T - t_0)\int_{t_0}^{t} E\,I_N(s)\,|f(s, X_s) - f(s, Y_s)|^2\,ds$$

$$+2 \int_{t_0}^t E I_N (s) |G (s, X_s) - G (s, Y_s)|^2 \, ds.$$

We now use condition (6.2.4) for the integrands and we obtain, for $L = 2 (T - t_0 + 1) K^2$,

(6.2.7) $\qquad E I_N (t) |X_t - Y_t|^2 \leq L \int_{t_0}^t E I_N (s) |X_s - Y_s|^2 \, ds.$

From this we would like to conclude that

(6.2.8) $\qquad E I_N (t) |X_t - Y_t|^2 = 0, \quad t \in [t_0, T],$

which we can, in fact, do with the aid of the following lemma of Bellman and Gronwall: If $g \geq 0$ and h are integrable on $[t_0, T]$ and if

$$g (t) \leq L \int_{t_0}^t g (s) \, ds + h (t), \quad t_0 \leq t \leq T,$$

for $L > 0$, then

$$g (t) \leq h (t) + L \int_{t_0}^t e^{L(t-s)} h (s) \, ds, \quad t_0 \leq t \leq T.$$

Proof can be found, for example, in Gikhman and Skorokhod [5], p. 393. The choices $h (t) \equiv 0$ and

$$g (t) = E I_N (t) |X_t - Y_t|^2$$

yield (6.2.8); that is,

$$I_N (t) X_t = I_N (t) Y_t \quad \text{with probability } 1$$

for every fixed $t \in [t_0, T]$. The inequality

$$P [I_N (t) \not\equiv 1 \text{ in } [t_0, T]] \leq P [\sup_{t_0 \leq t \leq T} |X_t| > N] + P [\sup_{t_0 \leq t \leq T} |Y_t| > N]$$

and the continuity (hence boundedness) of X_t and Y_t imply that we can make the right-hand member of the last inequality arbitrarily small by taking N sufficiently great. Therefore,

$$X_t = Y_t$$

with probability 1 for every fixed $t \in [t_0, T]$ and hence for a countable dense set M in $[t_0, T]$. Now X_t and Y_t are assumed to be almost certainly continuous, so that coincidence in M implies coincidence throughout the entire interval $[t_0, T]$ and hence

$$P\left[\sup_{t_0 \le t \le T} |X_t - Y_t| > 0\right] = 0.$$

This proves the uniqueness of a continuous solution—with only satisfaction of the Lipschitz condition (6.2.4) assumed.

b) Existence. We first treat the case

$$E\,|c|^2 < \infty.$$

Let us begin an iteration procedure with $X_t^{(0)} \equiv c$ and let us define, for $n \ge 1$ and $t \in [t_0, T]$,

$$(6.2.9) \qquad X_t^{(n)} = c + \int_{t_0}^{t} f\,(s, X_s^{(n-1)})\,\mathrm{d}s + \int_{t_0}^{t} G\,(s, X_s^{(n-1)})\,\mathrm{d}W_s.$$

If $X_t^{(n-1)}$ is nonanticipating and continuous, then, by virtue of the assumption (6.2.5), the right-hand member of (6.2.9) is a stochastic differential and hence, in accordance with remark (5.3.2), defines a nonanticipating continuous process X_t on $[t_0, T]$. Now, $X_t^{(0)}$ is nonanticipating and continuous, and with it so are all the processes $X_t^{(n)}$ for $n \ge 1$.

Let us now show that $X_t^{(n)}$ converges uniformly on $[t_0, T]$ to a solution X_t of equation (6.2.3) (in the sense of definition (6.1.3)). The assumption $E\,|c|^2 < \infty$ implies

$$\sup_{t_0 \le t \le T} E\,|X_t^{(0)}|^2 < \infty.$$

That this holds also for the subsequent processes $X_t^{(n)}$ can be seen as follows: by virtue of the inequality $|x + y + z|^2 \le 3\,(|x|^2 + |y|^2 + |z|^2)$ and (6.2.5), it follows from (6.2.9) that

$$E\,|X_t^{(n)}|^2 \le 3\,E\,|c|^2 + 3\,(T - t_0) \int_{t_0}^{t} K^2\,(1 + E\,|X_s^{(n-1)}|^2)\,\mathrm{d}s$$

$$+ 3 \int_{t_0}^{t} K^2\,(1 + E\,|X_s^{(n-1)}|^2)\,\mathrm{d}s$$

$$\le 3\,E\,|c|^2 + 3\,(T - t_0 + 1)\,K^2\,(T - t_0)\,\Bigl(1 + \sup_{t_0 \le t \le T} E\,|X_t^{(n-1)}|^2\Bigr),$$

so that

$$(6.2.10) \qquad \sup_{t_0 \le t \le T} E\,|X_t^{(n)}|^2 < \infty \qquad \text{for all} \qquad n \ge 1,$$

since this is true for $n = 0$.

Since $E |c|^2 < \infty$, equation (6.2.7) for $X_t^{(n+1)} - X_t^{(n)}$ can now be derived without $I_N(t)$; that is,

$$E |X_t^{(n+1)} - X_t^{(n)}|^2 \leq L \int_{t_0}^{t} E |X_s^{(n)} - X_s^{(n-1)}|^2 \, ds$$

with the same constant $L = 2 (T - t_0 + 1) K^2$. If we carry out an iteration on this inequality, using the familiar Cauchy formula

$$\int_{t_0}^{t} \int_{t_0}^{t_{n-1}} \cdots \int_{t_0}^{t_1} g(s) \, ds \, dt_1 \ldots dt_{n-1} = \int_{t_0}^{t} g(s) \frac{(t-s)^{n-1}}{(n-1)!} \, ds,$$

we get

$$(6.2.11) \qquad E |X_t^{(n+1)} - X_t^{(n)}|^2 \leq L^n \int_{t_0}^{t} \frac{(t-s)^{n-1}}{(n-1)!} E |X_s^{(1)} - X_s^{(0)}|^2 \, ds.$$

Now, under the assumption (6.2.5),

$$E |X_t^{(1)} - X_t^{(0)}|^2 \leq 2 (T - t_0 + 1) K^2 \int_{t_0}^{t} (1 + E |c|^2) \, ds$$

$$\leq L (T - t_0) (1 + E |c|^2) = C.$$

Therefore, it follows from (6.2.11) that

$$(6.2.12) \qquad \sup_{t_0 \leq t \leq T} E |X_t^{(n+1)} - X_t^{(n)}|^2 \leq C (L (T - t_0))^n / n!, \quad n \geq 0.$$

To prove the uniform convergence of $X_t^{(n)}$ itself, we need to find an estimate for

$$d_n = \sup_{t_0 \leq t \leq T} |X_t^{(n+1)} - X_t^{(n)}|$$

It follows from (6.2.9) that

$$d_n \leq \int_{t_0}^{T} |f(s, X_s^{(n)}) - f(s, X_s^{(n-1)})| \, ds$$

$$+ \sup_{t_0 \leq t \leq T} \left| \int_{t_0}^{t} (G(s, X_s^{(n)}) - G(s, X_s^{(n-1)})) \, dW_s \right|.$$

If we now use inequality (5.1.5) and the Lipschitz condition, we get

$$E d_n^2 \leq 2 (T - t_0) K^2 \int_{t_0}^{T} E |X_s^{(n)} - X_s^{(n-1)}|^2 \, ds + 2 \cdot 4 K^2 \int_{t_0}^{T}$$

$$E |X_s^{(n)} - X_s^{(n-1)}|^2 \, ds.$$

Then, with (6.2.12) for $n \geq 0$,

$$E \, d_n^2 \leq (2 \, (T - t_0) + 8) \, K^2 \, (T - t_0) \, C \, (L \, (T - t_0))^{n-1} / (n-1)!$$
$$= C_1 \, (L \, (T - t_0))^{n-1} / (n-1)!.$$

By virtue of the Borel-Cantelli lemma and Weierstrass's convergence criterion, convergence of the series

$$\sum_{n=1}^{\infty} P \, [d_n > n^{-2}] \leq C_1 \sum_{n=1}^{\infty} (L \, (T - t_0))^{n-1} \, n^4 / (n-1)!$$

implies that

$$\text{ac-}\lim_{n \to \infty} \left(X_t^{(0)} + \sum_{i=1}^{n} (X_t^{(i)} - X_t^{(i-1)}) \right) = \text{ac-}\lim_{n \to \infty} X_t^{(n)} = X_t$$

uniformly on $[t_0, T]$. Since X_t is the limit of a sequence of nonanticipating functions and the uniform limit of a sequence of continuous functions, it is itself nonanticipating and continuous. Because of the restriction on the increase in f and G and the continuity of X_t, the right-hand member of (6.2.3) becomes, when we substitute X_t into it, a meaningful stochastic differential. Therefore, it remains to show that X_t satisfies the equation

$$(6.2.13) \quad X_t = c + \int_{t_0}^{t} f \, (s, X_s) \, ds + \int_{t_0}^{t} G \, (s, X_s) \, ds$$

for all $t \in [t_0, T]$. Since $X_{t_0}^{(n)} = c$ (for $n \geq 0$), this is obvious for $t = t_0$. For $t \in (t_0, T]$ we take the limit in equation (6.2.9). By virtue of (6.2.4) and the uniform convergence of $X_t^{(n)}$, we have with probability 1

$$\left| \int_{t_0}^{t} f \, (s, X_s^{(n)}) \, ds - \int_{t_0}^{t} f \, (s, X_s) \, ds \right| \leq K \int_{0}^{t} |X_s^{(n)} - X_s| \, ds \to 0$$

and

$$\int_{t_0}^{t} |G \, (s, X_s^{(n)}) - G \, (s, X_s)|^2 \, ds \leq K^2 \int_{t_0}^{t} |X_s^{(n)} - X_s|^2 \, ds \to 0,$$

so that

$$\text{ac-}\lim_{n \to \infty} \int_{t_0}^{t} f \, (s, X_s^{(n)}) \, ds = \int_{t_0}^{t} f \, (s, X_s) \, ds$$

and, by Theorem (4.4.14d),

$$\text{st-}\lim_{n\to\infty} \int_{t_0}^{t} G\left(s, X_s^{(n)}\right) dW_s = \int_{t_0}^{t} G\left(s, X_s\right) dW_s .$$

Equation (6.2.13) then follows. Thus, X_t is a solution of equation of (6.2.3). The case of a general initial condition c is reduced to the case $E\,|c|^2 < \infty$ by defining

$$c_N = \begin{cases} c, & \text{if } |c| \leqq N, \\ 0, & \text{otherwise,} \end{cases}$$

and taking the limit as $N \to \infty$. For more details, see, for example, Gikhman and Skorokhod [5], pp. 395-397. ∎

6.3 Supplements to the Existence-and-Uniqueness Theorem

The method used for proving Theorem (6.2.2) consists basically in the Picard-Lindelöf iteration procedure with the Borel-Cantelli lemma forming the bridge from estimation of the error to convergence of the iteration procedure.

The Lipschitz condition (6.2.4) guarantees that $f\,(t,\,x)$ and $G\,(t,\,x)$ do not change faster with change in x than does the function x itself. This implies in particular the continuity of $f\,(t,\cdot)$ and $G\,(t,\cdot)$ for all $t \in [t_0,\,T]$. Thus, functions that are discontinuous with respect to x and even continuous functions of the type

$$f\,(t,\,x) = |x|^\alpha, \quad 0 < \alpha < 1,$$

are excluded as coefficients. It is known that the classical problem $(d = 1)$

$$X_t = \int_{t_0}^{t} |X_s|^\alpha \, ds$$

has, for $\alpha \geqq 1$, only the solution $X_t \equiv 0$ but, for $0 < \alpha < 1$, it has also the solution

$$X_t = (\beta\,(t - t_0))^{1/\beta}$$

where $\beta = 1 - \alpha$. I.V. Girsanov has shown [37] that the equation

$$X_t = \int_{0}^{t} |X_s|^\alpha \, dW_s \quad (d = m = 1)$$

has exactly one nonanticipating solution for $\alpha \geqq 1/2$ but infinitely many for $0 < \alpha < 1/2$.

In order to be able to admit functions like $\sin x^2$ (which become steeper and steeper with increasing x) as coefficients, we need

(6.3.1) Corollary. The existence-and-uniqueness theorem remains valid if we replace the Lipschitz condition with the more general condition that, for every $N > 0$, there exist a constant K_N such that, for all $t \in [t_0, T]$, $|x| \leq N$, and $|y| \leq N$,

$$(6.3.2) \qquad |f(t, x) - f(t, y)| + |G(t, x) - G(t, y)| \leq K_N |x - y|.$$

To prove this, we again use a truncation procedure, as we have done several times already, and then take the limit as $N \to \infty$. For more details, see Gikhman and Skorokhod [36], pp. 45-47.

A convenient sufficient condition for satisfaction of a Lipschitz condition follows from the mean-value theorem of differential calculus. According to this theorem, for a scalar function $g(x)$, where $x \in R^d$, with partial derivatives g_{x_i}, we have

$$g(b) - g(a) = g_x(a + \vartheta(b - a))'(b - a); \quad 0 < \vartheta < 1; \ a, b \in R^d.$$

where $g_x = (g_{x_1}, \dots, g_{x_d})'$. If $|g_x|$ is bounded on R^d by a constant C, then, for all x and y in R^d,

$$|g(y) - g(x)| \leq \sup_{z \in R^d} |g_x(z)| \, |y - x| \leq C |y - x|,$$

which is a Lipschitz condition. When we apply this to each component of f and G, we get

(6.3.3) Corollary. For the Lipschitz condition in the existence-and-uniqueness theorem (or its generalization (6.3.2)) to be satisfied, it is sufficient that the functions $f(t, x)$ and $G(t, x)$ have continuous partial derivatives of first order with respect to the components of x for every $t \in [t_0, T]$ and that these be bounded on $[t_0, T] \times R^d$ (or, in the case of the generalization, on $[t_0, T] \times \{|x| \leq N\}$).

Let us now discuss the meaning of the second assumption (namely, inequality (6.2.5)) in the existence-and-uniqueness theorem. This assumption bounds f and G uniformly with respect to $t \in [t_0, T]$ and allows at most linear increase of these functions with respect to x. If this condition is violated, we get the effect (familiar from the study of ordinary differential equations) of an "explosion" of the solution. Let us illustrate this with the scalar-ordinary-differential-equation initial-value problem

$$dX_t = X_t^2 \, dt, \quad X_0 = c.$$

The solution is

$$X_t = \begin{cases} 0, & \text{if } c = 0, \\ (1/c - t)^{-1}, & \text{if } c \neq 0. \end{cases}$$

Thus, the trajectory X_t is defined for $c > 0$ only on the interval $[0, 1/c)$. At $t =$

$1/c$, a so-called **explosion** takes place. For given $[0, T]$, there always exist initial values, namely, those for which $c \geqq 1/T$, for which the solution X_t is not defined throughout the entire interval $[0, T]$. The restriction on the growth of f and G guarantees that, with probability 1, the solution X_t does not explode in the interval $[t_0, T]$, whatever the initial value $X_{t_0} = c$. For further remarks about explosions, see McKean [45] and also (6.3.6)-(6.3.8).

(6.3.4) Remarks concerning global solutions. If the functions f and G are defined on $[t_0, \infty) \times R^d$ and if the assumptions of the existence-and-uniqueness theorem hold on *every* finite subinterval $[t_0, T]$ of $[t_0, \infty)$, then the equation

$$X_t = c + \int_{t_0}^{t} f(s, X_s) \, ds + \int_{t_0}^{t} G(s, X_s) \, dW_s,$$

has a unique solution X_t defined on the entire half-line $[t_0, \infty)$. Such a solution is called a **global** solution. The assumptions listed are satisfied, in particular, in the following special case:

(6.3.5) Corollary. Consider the *autonomous* stochastic differential equation

$$dX_t = f(X_t) \, dt + G(X_t) \, dW_t, \quad X_{t_0} = c,$$

(By autonomous is meant that $f(t, x) \equiv f(x)$ and $G(t, x) \equiv G(x)$, where $f(x) \in R^d$ and $G(x)$ is a $d \times m$ matrix.) For every initial value c that is independent of the m-dimensional Wiener process $W_t - W_{t_0}$ for $t \geqq t_0$, this equation has exactly one continuous global solution X_t on the entire interval $[t_0, \infty)$ such that $X_{t_0} = c$ provided only the following Lipschitz condition is satisfied: there exists a positive constant K such that, for all $x, y \in R^d$,

$$|f(x) - f(y)| + |G(x) - G(y)| \leqq K |x - y|.$$

The restriction on the growth of f and G follows from this global Lipschitz condition (we fix $y = y_0$).

(6.3.6) Example. Suppose that $d = m = 1$. Consider the autonomous equation

$$dX_t = -\frac{1}{2} \exp(-2 X_t) \, dt + \exp(-X_t) \, dW_t.$$

The coefficients in this equation do not satisfy any Lipschitz condition (or any growth restriction) for $x < 0$. Therefore, we must allow for possible explosions of the sample functions of X_t. The function

$$X_t = \log(W_t - W_{t_0} + e^c)$$

is a unique *local* solution on the interval $[t_0, \eta)$ with instant of explosion

$$\eta = \inf(t : W_t - W_{t_0} = -e^c) > 0$$

One can verify this formally with the aid of Itô's theorem. The existence and uniqueness are consequences of the following theorem of McKean ([45], p. 54):

(6.3.7) Theorem. Suppose that $d = m = 1$. The autonomous stochastic differential equation

$$dX_t = f(X_t)\, dt + G(X_t)\, dW_t, \quad X_{t_0} = c,$$

where f and G are continuously differentiable functions, has a unique local solution that is defined up to a (random) explosion time η in the interval $t_0 < \eta \leqq \infty$. If $\eta < \infty$, then $X_{\eta-0} = -\infty$ or $+\infty$.

(6.3.8) Remark. In accordance with Corollary (6.3.5), the satisfaction of a Lipschitz condition (in fact, the growth restriction) on f and G implies that

$$\eta = \infty \text{ with probability 1.}$$

A second sufficient condition for non-occurrence of an explosion in the case of continuously differentiable G is that $f \equiv 0$, that is, that the systematic part be absent. This follows from application of a sensitive test for explosion discovered by W. Feller. (This test, as well as a d-dimensional analogue discovered by Khas'-minskiy, can be found in McKean [45], p. 65.) It should be emphasized that, for $d \geqq 2$, the condition $f \equiv 0$ is no longer in general sufficient to preclude an explosion (see McKean [45], p. 106 (problem 3)) if the growth of G is not restricted.

(6.3.9) Remark. The above cited example of Girsanov

$$dX_t = |X_t|^\alpha\, dW_t, \quad d = m = 1, \quad c = 0,$$

suggests that the Lipschitz condition for G might possibly be replaced in the scalar autonomous case with the so-called Hölder condition with exponent $\alpha > 1/2$:

$$|G(x) - G(y)| \leqq K\,|x - y|^\alpha, \quad x, y \in R^1, \quad \alpha > 1/2, \quad K \geqq 0,$$

This is in fact the case (see, for example, W. J. Anderson [30], p. 76).

(6.3.10) Remark. By definition, the solution X_t of a stochastic differential equation (as well as $X_t - X_u$, where $t \geqq u \geqq t_0$) is nonanticipating and hence is, fc every $t \in [t_0, T]$, statistically independent of $W_s - W_t$, for $s \geqq t$. We emphasize once again the functional point of view and refer to remark (6.1.4). By virtue of the construction of the solution by means of an iteration procedure, X_t is not only $\mathfrak{F}_t$-measurable but measurable with respect to the sigma-algebra generated by c and $W_s - W_{t_0}$ for $t_0 \leqq s \leqq t$. The function g mentioned in remark (6.1.4) therefore depends only on c and $W_s - W_{t_0}$, for $t_0 \leqq s \leqq t$:

$$X_t(\omega) = g(c(\omega); W_s(\omega) - W_{t_0}(\omega), \quad t_0 \leqq s \leqq t).$$

For illustration, consider the example (6.3.12). More generally, we have the result that X_t is a function of X_s and $W_u - W_s$ for $s \leqq u \leqq t$, i.e.

$$X_t = g_s(X_s; W_u - W_s, \quad s \leqq u \leqq t).$$

(6.3.11) **Example.** Let us return to example (6.1.11) and consider, for $d = m = 1$, the stochastic differential equation

$$(6.3.12) \qquad dX_t = g(t) X_t dW_t, \quad X_{t_0} = c, \quad t_0 \leq t \leq T.$$

Thus, $f(t, x) \equiv 0$ and $G(t, x) = g(t) x$. The assumptions of the existence-and-uniqueness theorem are satisfied if $g(t)$ is measurable and bounded on $[t_0, T]$. Therefore a unique solution exists. We assert that this solution is

$$X_t = c \exp\left(-\frac{1}{2} \int_{t_0}^{t} g(s)^2 \, ds + \int_{t_0}^{t} g(s) \, dW_s\right).$$

In any case, $X_{t_0} = c$. If $c = 0$, then $X_t \equiv 0$ is obviously a solution. Let us now suppose that $c > 0$. If we set

$$Y_t = \log c - \frac{1}{2} \int_{t_0}^{t} g(s)^2 \, ds + \int_{t_0}^{t} g(s) \, dW_s$$

and evaluate the stochastic differential of $X_t = \exp Y_t$ with the aid of Itô's theorem, we get equation (6.3.12). If $c < 0$, we consider the process $-X_t$ and obtain the same result.

Other examples will be found, in particular, in Chapter 8.

Chapter 7

Properties of the Solutions of Stochastic Differential Equations

7.1 The Moments of the Solutions

In this section, we shall assume that the conditions of the existence-and-uniqueness theorem (6.2.2) are satisfied and we shall investigate the moments $E |X_t|^k$ of the solution of

$$(7.1.1) \qquad dX_t = f(t, X_t)\, dt + G(t, X_t)\, dW_t, \quad X_{t_0} = c,$$

on $[t_0, T]$. These moments do not in general have to exist but their existence carries over from the initial value c for all values X_t. More precisely, we have

(7.1.2) **Theorem.** Suppose that the assumptions of the existence-and-uniqueness theorem are satisfied and that

$$E |c|^{2n} < \infty$$

where n is a positive integer. Then, for the solution X_t of the stochastic differential equation (7.1.1) on $[t_0, T]$, where $T < \infty$,

$$(7.1.3) \qquad E |X_t|^{2n} \leqq (1 + E |c|^{2n})\, e^{C(t-t_0)}$$

and

$$(7.1.4) \qquad E |X_t - c|^{2n} \leqq D (1 + E |c|^{2n})(t - t_0)^n\, e^{C(t-t_0)},$$

where $C = 2n(2n+1)K^2$ and D are constants (dependent only on n, K, and $T - t_0$).

Proof. In accordance with Itô's theorem, $|X_t|^{2n}$ has a stochastic differential with the following integral form

$$|X_t|^{2n} = |c|^{2n} + \int_{t_0}^{t} 2n\, |X_s|^{2n-2}\, X_s'\, f(s, X_s)\, ds + \int_{t_0}^{t} 2n\, |X_s|^{2n-2}\, X_s'$$

$$G(s, X_s)\, dW_s + \int_{t_0}^{t} n\, |X_s|^{2n-2}\, |G(s, X_s)|^2\, ds$$

$$+ \int_{t_0}^t 2\,n\,(n-1)\,|X_s|^{2\,n-4}\,|X_s'\,G\,(s,X_s)|^2\,ds$$

(where the last term is absent if $n = 1$). That $E\,|X_t|^{2n}$ exists if $E\,|c|^{2n} < \infty$ follows from step b) in the proof of Theorem (6.2.2). We take the expectation on both sides of the last equation and, keeping the relationship

$$E\left(\int_{t_0}^t 2\,n\,|X_s|^{2\,n-2}\,X_s'\,G\,(s,X_s)\,dW_s\right) = 0$$

(see (4.5.14)) and equation (6.2.5) in mind, we obtain

$$E\,|X_t|^{2\,n} = E\,|c|^{2\,n} + \int_{t_0}^t E\,(2\,n\,|X_s|^{2\,n-2}\,X_s'\,f\,(s,X_s)$$

$$+ n\,|X_s|^{2\,n-2}|G\,(s,X_s)|^2 + 2\,n(n-1)\,|X_s|^{2\,n-4}\,|X_s'\,G\,(s,X_s)|^2)\,ds$$

$$\leqq E\,|c|^{2\,n} + (2\,n+1)\,n\,K^2 \int_{t_0}^t E\,(1+|X_s|^2)\,|X_s|^{2\,n-2}\,ds.$$

Since $(1+|x|^2)\,|x|^{2\,n-2} \leqq 1 + 2\,|x|^{2\,n}$, we have

$$E\,|X_t|^{2\,n} \leqq E\,|c|^{2\,n} + (2\,n+1)\,n\,K^2\,(t-t_0)$$

$$+ 2\,n\,(2\,n+1)\,K^2 \int_{t_0}^t E\,|X_s|^{2\,n}\,ds,$$

Therefore, by virtue of the Bellman-Gronwall lemma already used in the proof of Theorem (6.2.2), we have

$$E\,|X_t|^{2\,n} \leqq h\,(t) + 2\,n\,(2\,n+1)\,K^2 \int_{t_0}^t \exp\,(2\,n\,(2\,n+1)\,K^2\,(t-s))$$

$$h\,(s)\,ds$$

where

$$h\,(t) = E\,|c|^{2\,n} + (2\,n+1)\,n\,K^2\,(t-t_0),$$

from which (7.1.3) follows.

Inequality (7.1.4) is obtained in a similar manner. We shall treat only the case $n = 1$ and refer the reader to Gikhman and Skorokhod [36], pp. 49-50, for general n (though the scalar case).

Since $|a + b|^2 \leqq 2\,(|a| + |b|)$, we have

$$E\,|X_t - c|^2 \leqq 2\,E\left|\int_{t_0}^t f\,(s,X_s)\,ds\right|^2 + 2\,E\left|\int_{t_0}^t G\,(s,X_s)\,dW_s\right|^2$$

$$\leq 2 (T-t_0) \int_{t_0}^{t} E |f(s, X_s)|^2 \, ds + 2 \int_{t_0}^{t} E |G(s, X_s)|^2 \, ds.$$

The growth restriction (6.2.5) yields, with $L = 2 (T - t_0 + 1) K^2$,

$$E |X_t - c|^2 \leq L \int_{t_0}^{t} (1 + E |X_s|^2) \, ds,$$

and, with the result (7.1.3),

$$E |X_t - c|^2 \leq L \int_{t_0}^{t} (1 + (1 + E |c|^2) e^{C(s-t_0)}) \, ds$$

$$= L (t-t_0) (1 + (1 + E |c|^2) (e^{C(t-t_0)} - 1)/C (t-t_0))$$

$$\leq L (t-t_0) (1 + (1 + E |c|^2) e^{C(t-t_0)})$$

$$\leq D (1 + E |c|^2) (t-t_0) e^{C(t-t_0)}$$

with $D = 2 L$. ∎

(7.1.5) **Remark.** By virtue of remark (6.1.7), $X_t = X_t (t_0, c)$ is also a solution of the same stochastic differential equation on every subinterval $[s, T]$, where $t_0 \leq s$, with the initial condition $X_s = X_s (t_0, c)$. Therefore, in (7.1.4), we can replace c with X_s and t_0 with s. Then, by virtue of (7.1.3), we get the inequality

$$E |X_t - X_s|^{2n} \leq C_1 |t - s|^n, \quad t, s \in [t_0, T],$$

where C_1 depends only on $n, K, T - t_0$, and $E |c|^{2n}$. For $n = 1$, it then follows under the assumption $E |c|^2 < \infty$ that

$$\lim_{t \to s} E |X_t - X_s|^2 = 0,$$

that is, the solution X_t is mean-square-continuous at every point of the interval $[t_0, T]$ (but this does not imply mean-square differentiability [see also section 7.2]).

Of great importance are the functions $E X_t$ and $K(s, t) = E X_s X_t'$, which are meaningful for $E |c|^2 < \infty$ although these do not in the general (nonlinear) case satisfy any simple equation. For example,

$$m_t = E X_t = E c + \int_{t_0}^{t} E f(s, X_s) \, ds, \quad t_0 \leq t \leq T,$$

but $E f(s, X_s)$ cannot in general be expressed as a function of m_s. A similar situation (see example (5.4.9)) holds for

$$\operatorname{tr} K(t,t) = E\,|X_t|^2 = E\,|c|^2 + \int_{t_0}^{t} 2\,E\,(X_s';\, f\,(s, X_s))\,ds$$
$$+ \int_{t_0}^{t} E\,|G\,(s, X_s)|^2\,ds.$$

Closed expressions can be obtained for m_t and $K(s,t)$ in the linear case (see Chapter 8). For the case $d = 2$, $m = 1$, the following is an example:

(7.1.6) **Example** (of a second-order stochastic differential equation). J. Goldstein [37a] has investigated the scalar second-order differential equation

$$\ddot{Y}_t = f(t, Y_t, \dot{Y}_t) + G(t, Y_t, \dot{Y}_t)\,\xi_t, \quad t_0 \leq t \leq T,$$

with initial conditions $Y_{t_0} = c_0$, $\dot{Y}_{t_0} = c_1$, which is disturbed by a scalar white noise. Using remark (6.1.6), we convert this equation into a stochastic differential equation for the two-dimensional process $X_t = (Y_t, \dot{Y}_t)'$:

$$dX_t = d\begin{pmatrix} Y_t \\ \dot{Y}_t \end{pmatrix} = \begin{pmatrix} \dot{Y}_t \\ f(t, Y_t, \dot{Y}_t) \end{pmatrix} dt + \begin{pmatrix} 0 \\ G(t, Y_t, \dot{Y}_t) \end{pmatrix} dW_t.$$

The existence-and-uniqueness theorem (6.2.2) ensures the existence of a unique solution if the original coefficients f and G satisfy the Lipschitz and boundedness conditions.

The sample functions of the process Y_t are are differentiable with derivative $\dot{Y}_t$ although $\dot{Y}_t$ is not in general of bounded variation or differentiable (see section (7.2)). If Y_t is interpreted as the position of a particle, this particle possesses a velocity though in general (for $G \neq 0$) no acceleration. Compare Chapter 8 for linear f and G.

For the evaluation of the first two moments of Y_t and $\dot{Y}_t$ (whose existence is ensured, by virtue of (7.1.2), under the assumptions $E\,c_0^2 < \infty$, $E\,c_1^2 < \infty$, we write for simplicity $f_0(t) = f(t, Y_t, \dot{Y}_t)$ and $G_0(t) = G(t, Y_t, \dot{Y}_t)$. Then,

$$E\,Y_t = E\,c_0 + \int_{t_0}^{t} E\,\dot{Y}_s\,ds,$$

$$E\,\dot{Y}_t = E\,c_1 + \int_{t_0}^{t} E\,f_0(s)\,ds,$$

and, for the variances (see Goldstein [37a], pp. 48-49),

$$V(Y_t) = V(c_0) + V\left(t\,c_1 + \int_{t_0}^{t} (t-s)\,f_0(s)\,ds\right) + \int_{t_0}^{t} (t-s)^2\,E\,G_0(s)^2\,ds,$$

$$V(\dot{Y}_t) = V(c_1) + V\left(\int_{t_0}^{t} f_0(s)\,ds\right) + \int_{t_0}^{t} E\,G_0(s)^2\,ds.$$

In particular, if $|G(t, x^1, x^2)| \geq a > 0$ in $[t_0, T] \times R^2$, the last two equations imply

$$V(Y_t) \geq a^2 (t-t_0)^3/3,$$
$$V(\dot{Y}_t) \geq a^2 (t-t_0).$$

In the case $T = \infty$, the variances of both components must therefore approach ∞. A consequence of this is that neither Y_t nor $\dot{Y}_t$ can remain indefinitely in a bounded set.

7.2   Analytical Properties of the Solutions

Again, we assume that the conditions of the existence-and-uniqueness theorem (6.2.2) are satisfied. Let us investigate the sample functions of the solution process X_t of

$$(7.2.1) \qquad X_t = c + \int_{t_0}^{t} f(s, X_s)\, ds + \int_{t_0}^{t} G(s, X_s)\, dW_s$$

as R^d-valued functions of t on the interval $[t_0, T]$, where $T < \infty$.

We know from the existence-and-uniqueness theorem that almost all sample functions $X.(\omega)$ are continuous functions on $[t_0, T]$. The results already obtained regarding other properties of X_t allow us to state qualitatively that, as long as G does not vanish, the properties of W_t (unbounded variation, nondifferentiability, the local log log law [see section 3.1]) carry over to X_t. Here, the presence of the systematic term $f(t, X_t)\, dt$ plays no role since

$$X_t^{(1)} = \int_{t_0}^{t} f(s, X_s)\, ds$$

has absolutely continuous (that is, everywhere $[\lambda]$ differentiable and of bounded variation) and, in the case of continuous f, continuously differentiable sample functions such that

$$\dot{X}_t^{(1)} = f(t, X_t).$$

On the other hand, the fluctuational part

$$X_t^{(2)} = \int_{t_0}^{t} G(t, X_t)\, dW_t$$

reflects the above-mentioned irregularities no matter how smooth the function G may be as long as it does not vanish (see remark (5.2.8)). Only at those points such that $G(t, x) = 0$ can we hope for smooth (for example, differentiable) sample functions of X_t.

We now cite a few new results that provide a justification for these qualitative remarks.

(7.2.2) **Theorem** (Goldstein [37a], p. 31). If X_t is the solution of equation (7.2.1), $t_0 < t_1 < \ldots < t_n = T$ is a partition of the interval $[t_0, T]$, and $\delta_n = \max (t_k - t_{k-1})$, then

$$\text{st-}\lim_{\delta_n \to 0} \sum_{k=1}^{n} (X_{t_k} - X_{t_{k-1}})(X_{t_k} - X_{t_{k-1}})' = \int_{t_0}^{T} G(s, X_s) G(s, X_s)' \, ds,$$

and, in particular,

(7.2.3) $\quad \text{st-}\lim_{\delta_n \to 0} \sum_{k=1}^{n} |X_{t_k} - X_{t_{k-1}}|^2 = \int_{t_0}^{T} |G(s, X_s)|^2 \, ds.$

(7.2.4) **Corollary.** If, for some $p \in \{1, 2, \ldots, d\}$, the inequality

$$\sum_{j=1}^{m} |G_{pj}(t, X_t)|^2 > 0$$

holds at almost all $[\lambda]$ points $t \in [t_0, T]$ with probability 1, then the pth component of X_t is almost certainly of unbounded variation in every subinterval of $[t_0, T]$.

Proof. The assertions follow immediately from the inequality

$$\sum_{k=1} |X_{t_k}^p - X_{t_{k-1}}^p|^2 \leq \max_i |X_{t_i}^p - X_{t_{i-1}}^p| \sum_{k=1} |X_{t_k}^p - X_{t_{k-1}}^p|,$$

the continuity of X_t, and Theorem (7.2.2). ∎

With regard to the law of the iterated logarithm for X_t, one can find only results for the autonomous case in McKean [45], p. 96, and Anderson [30], pp. 51-57. However, Theorem 3.2.1 of Anderson ([30], p. 51) can be applied directly to the proof of the following assertion:

(7.2.5) **Theorem.** If X_t is the solution of equation (7.2.1) and if f and G are continuous with respect to t for $t \in [t_0, T]$, then, for every fixed $t \in [t_0, T]$,

$$\limsup_{h \to 0} \frac{|X_{t+h} - X_t|}{\sqrt{2h \log \log 1/h}} = \sqrt{\lambda(t, X_t)},$$

with probability 1, where $\lambda(t, x)$ is the greatest eigenvalue of the (necessarily nonnegative-definite) matrix $G(t, x) G(t, x)'$.

More precisely, we can show that the cluster points of $(X_{t+h} - X_t)/(2 h \log \log 1/h)^{1/2}$ as $h \to 0$ are almost certainly all points of that ellipsoid in R^d whose principal axes have the directions of the eigenvectors and whose lengths are the roots of the corresponding eigenvalues of $G(t, X_t) G(t, X_t)'$. Therefore, this ellipsoid depends on t and X_t (and hence also on ω) (see Arnold [30a]).

Just as in section 3.1 for W_t we can again conclude from the last remark that $X_{\cdot}(\omega)$ is nondifferentiable at the instant t provided $G(t, X_t(\omega)) \neq 0$.

The possibility of smooth behavior on the part of X_t occurs only in the case $G(t, X_t) = 0$. In this connection, we cite a result of Anderson ([30], p. 59):

(7.2.6)   **Theorem.** If X_t is the solution of equation (7.2.1), if f and G are continuous for $t \in [t_0, T]$, and if the initial value c is almost certainly constant, then, in the case $G(t_0, c) = 0$,

$$\lim_{t \to t_0} \frac{X_t - c}{t} = f(t_0, c).$$

with probability 1. Thus, differentiability of the solution of a stochastic differential equation is the exception and nondifferentiability is the rule. The formal differential equation

$$\dot{X}_t = f(t, X_t) + G(t, X_t) \xi_t,$$

where ξ_t is an m-dimensional white noise, cannot therefore as a rule be interpreted as an ordinary differential equation for the function X_t.

7.3   Dependence of the Solutions on Parameters and Initial Values

The value $X_t(\omega)$ of the solution trajectory of the stochastic differential equation

$$dX_t = f(t, X_t)\,dt + G(t, X_t)\,dW_t, \quad X_{t_0} = c, \quad t_0 \leq t \leq T,$$

is, in accordance with remark (6.3.10), a function (determined uniquely by f and G) of the two (independent) random elements $c(\omega)$ and $W_s(\omega) - W_{t_0}(\omega)$, for $t_0 \leq s \leq t$:

$$X_t(\omega) = g(c(\omega); W_s(\omega) - W_{t_0}(\omega), \quad t_0 \leq s \leq t).$$

Just as with ordinary differential equations, we are interested again here in the manner in which the function g depends on the initial value c and on any parameters that may appear in f and G.

In this section, we shall give two theorems, for the proof of which we refer to Gikhman and Skorokhod [36] or Skorokhod [47].

(7.3.1)   **Theorem.** Let $X_t(p)$ denote the solution of the stochastic differential equation

$$dX_t = f(p, t, X_t)\,dt + G(p, t, X_t)\,dW_t, \quad X_{t_0} = c(p),$$

in the interval $[t_0, T]$, for $T < \infty$. Here, p is a parameter, and the functions $f(p, t, x)$ and $G(p, t, x)$ satisfy, for all p, the conditions of the existence-and-uniqueness theorem. Suppose also that the following conditions are satisfied:

a) $$\text{st-}\lim_{p \to p_0} c(p) = c(p_0),$$

b) for every $N > 0$,

$$\lim_{p \to p_0} \sup_{t \in [t_0, T], |x| \leq N} (|f(p, t, x) - f(p_0, t, x)|$$
$$+ |G(p, t, x) - G(p_0, t, x)|) = 0,$$

c) there exists a constant K independent of p such that

$$|f(p, t, x)|^2 + |G(p, t, x)|^2 \leq K^2 (1 + |x|^2).$$

Then,

$$\text{st-}\lim_{p \to p_0} \sup_{t_0 \leq t \leq T} |X_t(p) - X_t(p_0)| = 0.$$

(7.3.2) **Remark.** If the functions f and G are independent of p, Theorem (7.3.1) implies the stochastically continuous dependence of the solution of a stochastic differential equation on the initial value c.

(7.3.3) **Examples.** a) Suppose that

$$dX_t(\varepsilon) = \varepsilon f(t, X_t(\varepsilon)) \, dt + dW_t,$$

where ε is a small parameter. The solution for $\varepsilon = 0$ is

$$X_t(0) = c(0) + W_t - W_{t_0}.$$

If

$$\text{st-}\lim_{\varepsilon \to 0} c(\varepsilon) = c(0)$$

we also have

$$\text{st-}\lim_{\varepsilon \to 0} \sup_{t_0 \leq t \leq T} |X_t(\varepsilon) - X_t(0)| = 0.$$

b) An analogous conclusion may be drawn in the case

$$dX_t(\varepsilon) = f(t, X_t(\varepsilon)) \, dt + \varepsilon \, dW_t,$$

where $X_t(0)$ is the solution of the ordinary differential equation $\dot{X}_t = f(t, X_t)$ with a possibly random initial value.

We shall now investigate the special case of a constant initial value though at an arbitrary instant $s \in [t_0, t]$. Let $X_t(s, x)$ denote the solution of the equation

(7.3.4) $$X_t(s, x) = x + \int_s^t f(u, X_u(s, x)) \, du + \int_s^t G(u, X_u(s, x)) \, dW_u,$$

where $t_0 \leq s \leq t \leq T$, which satisfies the initial condition $X_s(s, x) = x \in R^d$.

(7.3.5)   **Definition.** A stochastic process X_t of the real parameter t is said to be **mean-square-differentiable** at the point t_1 with random variable Y_{t_1} as its derivative if the second moments of X_t and Y_{t_1} exist and if

$$\lim_{h \to 0} E |(X_{t_1+h} - X_{t_1})/h - Y_{t_1}|^2 = 0.$$

This concept of a derivative is used when one investigates the dependence of the solution $X_t(s, x)$ of equation (7.3.4) on the parameter x.

(7.3.6)   **Theorem.** Suppose that the coefficients $f(t, x)$ and $G(t, x)$ of equation (7.3.4) are continuous with respect to (t, x) and that they have bounded continuous first and second partial derivatives with respect to the x_i, where $x = (x_1, \ldots, x_d)'$. Then, for fixed $t \in [s, T]$, the solution $X_t(s, x)$ is mean-square-continuous with respect to (s, x) and is twice mean-square-differentiable with respect to the x_i. The derivatives

$$\frac{\partial}{\partial x_i} X_t(s, x), \quad \frac{\partial^2}{\partial x_i \, \partial x_j} X_t(s, x)$$

are, as functions of x, mean-square-continuous and they satisfy the stochastic differential equations that one obtains from (7.3.4) by partial differentiation, so that, for example, for $Y_t = \partial X_t(s, x)/\partial x_i$,

$$Y_t = e_i + \int_s^t f_x(u, X_u(s, x)) Y_u \, du + \int_s^t G_x(u, X_u(s, x)) Y_u \, dW_u.$$

Here, $e_i \in R^d$ is the unit vector in the x_i direction, f_x is the $d \times d$ matrix with column vectors f_{x_j}, and $G_x = (G_{x_1}, \ldots, G_{x_d})$, where the G_{x_j} are $d \times m$ matrices.

(7.3.7)   **Remark.** For later use, we assert that, if the conditions of Theorem (7.3.6) are satisfied and if $g(x)$ is a bounded continuous function in R^d with bounded continuous first and second partial derivatives, then, for fixed $t \in [t_0, T]$, the function

$$u(s, x) = E g(X_t(s, x))$$

is continuous with respect to $(s, x) \in [t_0, t] \times R^d$, is bounded, and has continuous bounded first and second partial derivatives with respect to the x_i and a continuous bounded derivative with respect to s.

Chapter 8
Linear Stochastic Differential Equations

8.1 Introduction

For the differential equations of the type that we are studying in this book, namely,

$$\dot{X}_t = f(t, X_t) + G(t, X_t)\,\xi_t,$$

where ξ_t is a Gaussian white noise, the right-hand member represents a *linear* function of the disturbance ξ_t. On the other hand, the functions f and G are in general nonlinear functions of the state X_t of the system.

Just as with ordinary differential equations, a much more complete theory can be developed in the stochastic case when the coefficient functions $f(t, x)$ and $G(t, x)$ are linear functions of x, especially when G is independent of x.

(8.1.1) Definition. A stochastic differential equation

$$dX_t = f(t, X_t)\,dt + G(t, X_t)\,dW_t$$

for the d-dimensional process X_t on the interval $[t_0, T]$ is said to be **linear** if the functions $f(t, x)$ and $G(t, x)$ are linear functions of $x \in R^d$ on $[t_0, T] \times R^d$, in other words, if

$$f(t, x) = A(t)\,x + a(t),$$

where $A(t)$ is $(d \times d)$-matrix-valued and $a(t)$ is R^d-valued, and if

$$G(t, x) = (B_1(t)\,x + b_1(t), \dots, B_m(t)\,x + b_m(t)),$$

where $B_k(t)$ is $(d \times d)$-matrix-valued and $b_k(t)$ is R^d-valued. Thus, a linear differential equation has the form

$$dX_t = (A(t)\,X_t + a(t))\,dt + \sum_{i=1}^{m}(B_i(t)\,X_t + b_i(t))\,dW_t^i,$$

where $W_t = (W_t^1, \dots, W_t^m)'$. It is said to be **homogeneous** if $a(t) = b_1(t) = \dots = b_m(t) \equiv 0$. It is said to be **linear in the narrow sense** if $B_1(t) = \dots = B_m(t) \equiv 0$.

(8.1.2) **Remark.** We obtain linear stochastic differential equations if in ordinary linear differential equations of the form

$$\dot{X}_t = A(t) X_t + a(t)$$

the coefficients $A(t)$ and $a(t)$ are "noisy" and/or a disturbance term independent of the state X_t of the system is added to the right-hand member of this equation. Then, the equation takes the form

(8.1.3) $$\dot{X}_t = (A(t) + \Delta(t)) X_t + (a(t) + \alpha(t)).$$

Here, $\Delta(t)$ is a $d \times d$ matrix and $\alpha(t)$ is a d-dimensional vector whose elements are (possibly) correlated Gaussian noise processes with time-varying intensity such that

$$E\, \Delta_{ij}(t)\, \Delta_{kp}(s) = C_{ij,kp}(t)\, \delta(t-s),$$

$$E\, \Delta_{ij}(t)\, \alpha_k(s) = D_{ij,k}(t)\, \delta(t-s)$$

and

$$E\, \alpha_i(t)\, \alpha_k(s) = E_{ik}(t)\, \delta(t-s).$$

As usual, we make the substitution

$$(\Delta(t), \alpha(t))\, dt = dY_t$$

(the elements of $\Delta(t)$ and $\alpha(t)$ are assumed written columnwise successively as a $(d^2 + d)$-dimensional column vector), where Y_t is an m-dimensional Gaussian process with independent increments such that $m = d^2 + d$ and $E\, Y_t = 0$ and with $m \times m$ covariance matrix

$$E\, Y_t\, Y'_s = \int_{t_0}^{\min(t,s)} Q(u)\, du$$

where

$$Q(t) = \begin{pmatrix} C(t) & D(t) \\ D'(t) & E(t) \end{pmatrix}.$$

By virtue of remark (5.2.4), such a process (we assume integrability of $Q(t)$ over the interval $[t_0, T]$ can be represented as

$$Y_t = \int_{t_0}^{t} \sqrt{Q(s)}\, dW_s,$$

that is,

$$dY_t = \sqrt{Q(s)}\, dW_s,$$

where W_t is now an m-dimensional Wiener process (with *independent* components). Instead of $\sqrt{Q}$, any $m \times p$ matrix G such that $G\,G' = Q$ can be used. This is important if Q is a singular matrix because we can then choose $p < m$. If we decompose $\sqrt{Q}$ in the form

$$\sqrt{Q} = \begin{pmatrix} \bar{B}_1(t) \dots \bar{B}_m(t) \\ b_1(t) \; \dots \; b_m(t) \end{pmatrix}, \quad b_k \in R^d, \quad \bar{B}_k \in R^{d^2},$$

equation (8.1.3) acquires the desired differential form

$$dX_t = (A(t)\,X_t + a(t))\,dt + \sum_{i=1}^{m} (B_i(t)\,X_t + b_i(t))\,dW_t^i$$

where

$$B_i(t) = (B_i^{kp}(t)) = (\bar{B}_i^{pd+k}), \quad 1 \le k,\; p \le d.$$

Thus, B_i is the vector $\bar{B}_i$ arranged as a $d \times d$ matrix.

(8.1.4) **Example.** On the basis of remark (6.1.6), let us rewrite the nth-order scalar differential equation

$$Y_t^{(n)} + (b_1(t) + \xi_1(t))\,Y_t^{(n-1)} + \dots + (b_n(t) + \xi_n(t))\,Y_t$$
$$+ (b_{n+1}(t) + \xi_{n+1}(t)) = 0,$$

where the $\xi_i(t)$ are in general correlated Gaussian noise processes with covariance

$$E\,\xi_i(t)\,\xi_j(s) = Q_{ij}(t)\,\delta(t-s)$$

as a first-order differential equation for

$$X_t = \begin{pmatrix} X_t^1 \\ \vdots \\ X_t^n \end{pmatrix} = \begin{pmatrix} Y_t \\ \dot{Y}_t \\ \vdots \\ Y_t^{(n-1)} \end{pmatrix}, \quad d = n,$$

specifically,

$$\dot{X}_t^i = X_t^{i+1}, \quad i = 1, \dots, n-1,$$

$$\dot{X}_t^n = -\sum_{k=1}^{n} (b_k(t) + \xi_k(t))\,X_t^{n+1-k} - (b_{n+1}(t) + \xi_{n+1}(t)),$$

and finally, in accordance with (8.1.2), as a linear stochastic differential equation in X_t of the form

$$dX_t^i = X_t^{i+1}\,dt, \quad i = 1, \dots, n-1,$$

$$dX_t^n = \left(-\sum_{k=1}^{n} b_k(t) X_t^{n+1-k} - b_{n+1}(t) \right) dt$$

$$-\sum_{p=1}^{n+1} \left(\sum_{k=1}^{n} G_{kp}(t) X_t^{n+1-k} + G_{n+1,p}(t) \right) dW_t^p,$$

where G is an $(n+1)\times(n+1)$ matrix such that $G\,G' = Q$; hence, $m = n+1$.

The following theorem is an immediate consequence of the existence-and-uniqueness Theorem (6.2.2).

(8.1.5) **Theorem.** The linear stochastic differential equation

$$dX_t = (A(t) X_t + a(t))\, dt + \sum_{i=1}^{m} (B_i(t) X_t + b_i(t))\, dW_t^i$$

has, for every initial value $X_{t_0} = c$ that is independent of $W_t - W_{t_0}$ (where $t \geqq t_0$), a unique continuous solution throughout the interval $[t_0, T]$ provided only the functions $A(t)$, $a(t)$, $B_i(t)$, and $b_i(t)$ are measurable and bounded on that interval. If this assumption holds in every subinterval of $[t_0, \infty)$, there exists a unique global solution (i.e. defined for all $t \in [t_0, \infty)$).

(8.1.6) **Corollary.** A global solution always exists for the *autonomous* linear differential equation

$$dX_t = (A X_t + a)\, dt + \sum_{i=1}^{m} (B_i X_t + b_i)\, dW_t^i, \quad X_{t_0} = c,$$

(with coefficients A, a, B_i, and b_i independent of t).

We now wish, if possible, to get a closed and explicit expression for this solution and to investigate it.

8.2 Linear Equations in the Narrow Sense

In this section, we shall investigate those equations that are obtained from a deterministic linear system

$$\dot{X}_t = A(t) X_t + a(t),$$

(where $A(t)$ is a $d \times d$ matrix and X_t and $a(t)$ are vectors with components in R^d) by the addition of a fluctuational term

$$B(t) \xi_t,$$

(where $B(t)$ is a $d \times m$ matrix and ξ_t is an m-dimensional white noise) that is independent of the state of the system; that is, we shall investigate equations of the form

(8.2.1) $dX_t = (A(t) X_t + a(t)) dt + B(t) dW_t$.

Here, we have combined the m vectors b_i appearing in definition (8.1.1) into a single $d \times m$ matrix $B = (b_1, \ldots, b_m)$. If the functions $A(t), a(t)$, and $B(t)$ are measurable and bounded on $[t_0, T]$ (as we shall assume to be the case in what follows), there exists, by virtue of Theorem (8.1.5), for every initial value $X_{t_0} = c$ a unique solution.

Let us review a few familiar items regarding deterministic linear systems $(B(t) \equiv 0)$ (see, for example, Bucy and Joseph [61], p. 5).

The matrix $\Phi(t) = \Phi(t, t_0)$ of solutions of the homogeneous equation

$$\dot{X}_t = A(t) X_t$$

with unit vectors $c = e_i$ in the x_i-direction as initial value, in other words, the solution of the matrix equation

$$\dot{\Phi}(t) = A(t) \Phi(t), \quad \Phi(t_0) = I,$$

is called the **fundamental matrix** of the system

$$\dot{X}_t = A(t) X_t + a(t).$$

The solution with initial value $X_{t_0} = c$ can be represented with the aid of $\Phi(t)$ in the following form:

$$X_t = \Phi(t) \left(c + \int_{t_0}^{t} \Phi(s)^{-1} a(s) ds \right).$$

If, for example, $A(t) \equiv A$ is independent of t, then

$$\Phi(t) = e^{A(t-t_0)} = \sum_{n=0}^{\infty} A^n (t-t_0)^n / n!.$$

Therefore,

$$X_t = e^{A(t-t_0)} c + \int_{t_0}^{t} e^{A(t-s)} a(s) ds.$$

With this knowledge, we can now easily determine the solution of the "nonhomogeneous" equation (8.2.1):

(8.2.2) **Theorem.** The linear (in the narrow sense) stochastic differential equation

$$dX_t = (A(t) X_t + a(t)) dt + B(t) dW_t, \quad X_{t_0} = c,$$

has on $[t_0, T]$ the solution

(8.2.3) $$X_t = \Phi(t)\left(c + \int\limits_{t_0}^{t} \Phi(s)^{-1} a(s)\, ds + \int\limits_{t_0}^{t} \Phi(s)^{-1} B(s)\, dW_s\right).$$

Here, $\Phi(t)$ is the fundamental matrix of the deterministic equation $\dot{X}_t = A(t) X_t$.

Proof.   If we set

$$Y_t = c + \int\limits_{t_0}^{t} \Phi(s)^{-1} a(s)\, ds + \int\limits_{t_0}^{t} \Phi(s)^{-1} B(s)\, dW_s,$$

Y_t has the stochastic differential

$$dY_t = \Phi(t)^{-1}(a(t)\, dt + B(t)\, dW_t).$$

Then, in accordance with Theorem (5.3.8), the process

$$X_t = \Phi(t) Y_t$$

has the stochastic differential

$$\begin{aligned}
dX_t &= \dot{\Phi}(t) Y_t\, dt + \Phi(t)\, dY_t \\
&= A(t) \Phi(t) Y_t\, dt + a(t)\, dt + B(t)\, dW_t \\
&= (A(t) X_t + a(t))\, dt + B(t)\, dW_t. \quad\blacksquare
\end{aligned}$$

In the above representation of the solution, we see clearly that the value X_t is a functional (uniquely determined by the coefficients $A(t), a(t)$, and $B(t)$) of c and $W_s - W_{t_0}$ for $t_0 \leq s \leq t$ (see remark (6.3.10)).

We mention in particular the following special cases:

(8.2.4) **Corollary.** If the matrix $A(t) \equiv A$ in equation (8.2.1) is independent of t, then

$$X_t = e^{A(t-t_0)} c + \int\limits_{t_0}^{t} e^{A(t-s)}(a(s)\, ds + B(s)\, dW_s).$$

(8.2.5) **Corollary.** For $d = 1$ (but m arbitrary),

$$\Phi(t) = \exp\left(\int\limits_{t_0}^{t} A(s)\, ds\right),$$

and hence

$$X_t = \exp\left(\int\limits_{t_0}^{t} A(s)\, ds\right)\left(c + \int\limits_{t_0}^{t} \exp\left(-\int\limits_{t_0}^{s} A(u)\, du\right)(a(s)\, ds + B(s)\, dW_s)\right).$$

In accordance with (7.1.2), the solution X_t has moments of second order if $E\,|c|^2 < \infty$. In our special case, the first two moments of X_t can easily be calculated from the explicit form of the solution:

(8.2.6) **Theorem.** For the solution X_t of the linear stochastic differential equation

$$dX_t = (A\,(t)\,X_t + a\,(t))\,dt + B\,(t)\,dW_t, \quad X_{t_0} = c,$$

we have, under the assumption $E\,|c|^2 < \infty$,

a)
$$m_t = E\,X_t = \Phi\,(t)\left(E\,c + \int_{t_0}^{t} \Phi\,(s)^{-1}\,a\,(s)\,ds\right),$$

Therefore, m_t is the solution of the deterministic linear differential equation

$$\dot{m}_t = A\,(t)\,m_t + a\,(t), \quad m_{t_0} = E\,c.$$

b)
$$K\,(s, t) = E\,(X_s - E\,X_s)\,(X_t - E\,X_t)'$$

$$= \Phi\,(s)\left(E\,(c - E\,c)\,(c - E\,c)' + \int_{t_0}^{\min\,(s,\,t)} \Phi\,(u)^{-1}\,B\,(u)\right.$$

(8.2.7)
$$\left. B\,(u)'\,(\Phi\,(u)^{-1})'\,du\right)\Phi\,(t)'.$$

In particular, the covariance matrix of the components of X_t

$$K\,(t) = K\,(t, t) = E\,(X_t - E\,X_t)\,(X_t - E\,X_t)'$$

is the unique symmetric nonnegative-definite solution of the matrix equation

(8.2.8) $\qquad \dot{K}\,(t) = A\,(t)\,K\,(t) + K\,(t)\,A\,(t)' + B\,(t)\,B\,(t)'$

with the initial value $K\,(t_0) = E\,(c - E\,c)\,(c - E\,c)'$.

Proof. a) If we take the expectation on both sides of (8.2.3), we get the formula for m_t. Differentiation of this expression with respect to t yields

$$\dot{m}_t = A\,(t)\,m_t + a\,(t),$$

which we also get immediately from the (integral form of the) stochastic differential equation by taking the expectation.

b) The formula for $K\,(s, t)$ follows also from (8.2.3) if we keep in mind the independence of c and

$$\int_{t_0}^{t} \Phi\,(s)^{-1}\,B\,(s)\,dW_s\,.$$

In particular, if we differentiate $K\,(t, t)$ with respect to t, we get the differential equation for K which we could have obtained directly from the stochastic differential equation. The differential equation for $K(t) = K(t)'$ satisfies the Lipschitz and boundedness conditions on $[t_0, T]$, so that a unique solution exists. ∎

Equation (8.2.8) therefore represents (in view of the symmetry of K) a system of $d(d+1)/2$ linear equations.

(8.2.9) **Remark.** Of particular interest is the behavior of

$$E|X_t - EX_t|^2 = E|X_t|^2 - |m_t|^2 = \text{tr } K(t) = \sum_{i=1}^{d} K_{ii}(t).$$

By using the relationship $\text{tr } A A' = |A|^2$, we obtain from the formula for $K(s, t)$

$$\text{tr } K(t) = E|\Phi(t)(c - Ec)|^2 + \int_{t_0}^{t} |\Phi(t)\Phi(s)^{-1}B(s)|^2 \, ds.$$

In formula (8.2.3), the solution X_t is represented as the sum of three statistically independent terms, the second of which is completely independent of ω and the third is, in accordance with Corollary (4.5.6) normally distributed. Therefore, the process Y_t in

$$X_t = \Phi(t)c + Y_t$$

is always a Gaussian process independent of $\Phi(t)c$ with independent increments and with distribution

$$\mathfrak{N}\left(\Phi(t)\int_{t_0}^{t} \Phi(s)^{-1} a(s) \, ds, \int_{t_0}^{t} \Phi(t)\Phi(s)^{-1}B(s)B(s)'(\Phi(s)^{-1})'\Phi(t)' ds\right).$$

The process X_t is itself Gaussian if and only if the initial value c is normally distributed (or constant). We write this important special case as

(8.2.10) **Theorem.** The solution (8.2.3) of the linear equation

$$dX_t = (A(t)X_t + a(t)) \, dt + B(t) \, dW_t, \quad X_{t_0} = c,$$

is a *Gaussian* stochastic process X_t if and only if c is normally distributed or constant. The mean value m_t and the covariance matrix $E(X_s - m_s)(X_t - m_t)'$ are given in Theorem (8.2.6). The process X_t has independent increments if and only if c is constant or $A(t) \equiv 0$ (that is, $\Phi(t) \equiv I$).

Now that we know the process is Gaussian in the case of normally distributed c, the question arises as to when it is *stationary*. A necessary and sufficient condition for this is

$$m_t = \text{const},$$

$$K(s, t) = \overline{K}(s - t).$$

These conditions are certainly satisfied if

$$E c = 0,$$

$$a(t) \equiv 0$$

(in this case, $m_t \equiv 0$) and

$$A(t) \equiv A$$

$$B(t) \equiv B$$

(that is, the original equation is autonomous and the solution X_t exists on $[t_0, \infty)$). Furthermore, by virtue of (8.2.7),

(8.2.11) $$A \overline{K}(0) + \overline{K}(0) A' = - B B'$$

and

$$\overline{K}(0) = E c c'.$$

The matrix equation (8.2.11) has a nonnegative-definite solution $\overline{K}(0)$, namely,

$$\overline{K}(0) = \int_0^\infty e^{At} B B' e^{A't} \, dt,$$

if the deterministic equation $\dot{X}_t = A X_t$ is asymptotically stable (that is, if all the eigenvalues of A have negative real parts [see example (11.1.4) and Bucy and Joseph [61], p. 9]). Furthermore, from formula (8.2.7) we get for $t = s$

$$e^{-A(t-t_0)} \overline{K}(0) e^{-A'(t-t_0)} = \overline{K}(0) + \int_{t_0}^t e^{-A(s-t_0)} B B' e^{-A'(s-t_0)} \, ds.$$

Therefore, for general $t, s \geq t_0$,

$$\overline{K}(s-t) = K(s,t) = \begin{cases} e^{A(s-t)} \overline{K}(0), & s \geq t, \\ \overline{K}(0) e^{A'(t-s)}, & s \leq t. \end{cases}$$

We write this result as

(8.2.12) **Theorem.** The solution of the equation

$$dX_t = (A(t) X_t + a(t)) \, dt + B(t) \, dW_t, \quad X_{t_0} = c,$$

is a *stationary Gaussian process* if $A(t) \equiv A$, $a(t) \equiv 0$, $B(t) \equiv B$, the eigenvalues of A have negative real parts, and c is $\mathfrak{N}(0, K)$-distributed, where K is the solution

$$K = \int_0^\infty e^{At} B B' e^{A't} \, dt$$

of the equation $A K + K A' = -B B'$. Then, for the process X_t,

$$E X_t \equiv 0$$

and

$$E X_s X_t' = \begin{cases} e^{A(s-t)} K, & s \geq t \geq t_0, \\ K e^{A'(s-t)}, & t \geq s \geq t_0. \end{cases}$$

Obviously, under the above conditions, the process X_t is stationary in the wide sense with the above first and second moments even when c is not normally distributed but $E c = 0$ and $E c c' = K$.

8.3 The Ornstein-Uhlenbeck-Process

We shall now investigate the historically oldest example of a stochastic differential equation. For the Brownian motion of a particle under the influence of friction but no other force field, the so-called Langevin equation

$$\dot{X}_t = -\alpha X_t + \sigma \xi_t,$$

where $\alpha > 0$ and σ are constants, has been derived in many ways (see, for example, Uhlenbeck and Ornstein [49] or Wang and Uhlenbeck [50]). Here, X_t is one of the three scalar velocity components of the particle and ξ_t is a scalar white noise. The corresponding stochastic differential equation (here, $d = m = 1$ and we set $t_0 = 0$)

$$dX_t = -\alpha X_t \, dt + \sigma \, dW_t, \quad X_0 = c,$$

is linear in the narrow sense and autonomous. Therefore, in accordance with Corollary (8.2.4), its unique solution is

$$X_t = e^{-\alpha t} c + \sigma \int_0^t e^{-\alpha(t-s)} \, dW_s.$$

In accordance with (8.2.6), X_t has, in the case $E c^2 < \infty$, mean value

$$m_t = E X_t = e^{-\alpha t} E c$$

and covariance

$$K(s,t) = E(X_s - m_s)(X_t - m_t) = e^{-\alpha(t+s)} (\text{Var}(c) + \sigma^2 (e^{2\alpha \min(t,s)} - 1)/2 \alpha).$$

In particular,

$$K(t,t) = \text{Var}(X_t) = e^{-2\alpha t} \text{Var}(c) + \sigma^2 (1 - e^{-2\alpha t})/2 \alpha.$$

For arbitrary c,

$$\text{ac-}\lim_{t \to \infty} e^{-\alpha t} c = 0,$$

so that the distribution of X_t approaches $\mathfrak{N}(0, \sigma^2/2\,\alpha)$ for arbitrary c as $t \to \infty$. For normally distributed or constant c, the solution X_t is a Gaussian process, the so-called **Ornstein-Uhlenbeck velocity process**. If we begin with an $\mathfrak{N}(0, \sigma^2/2\,\alpha)$-distributed c, then X_t is a stationary Gaussian process (sometimes called a **colored noise**) such that $E\,X_t \equiv 0$ and

$$E\,X_s X_t = e^{-\alpha|t-s|}\,\sigma^2/2\,\alpha,$$

which we can also get from Theorem (8.2.12).

By integration of the velocity X_t, we obtain the position

$$Y_t = Y_0 + \int_0^t X_s \, ds$$

of the particle. If c and Y_0 are normally distributed or constant, Y_t is, with X_t, a Gaussian process, the so-called **Ornstein-Uhlenbeck (position) process**. Of course, we can treat X_t and Y_t simultaneously by combining their equations into the single equation

$$d\begin{pmatrix} X_t \\ Y_t \end{pmatrix} = \begin{pmatrix} -\alpha & 0 \\ 1 & 0 \end{pmatrix} \begin{pmatrix} X_t \\ Y_t \end{pmatrix} dt + \begin{pmatrix} \sigma \\ 0 \end{pmatrix} dW_t, \quad \begin{pmatrix} X_0 \\ Y_0 \end{pmatrix} = \begin{pmatrix} c \\ Y_0 \end{pmatrix}.$$

We have

$$E\,Y_t = E\,Y_0 + (1 - e^{-\alpha t})\,E\,c/\alpha$$

and

$$E\,(Y_s - E\,Y_s)\,(Y_t - E\,Y_t) = \text{Var}\,(Y_0) + 2\,D\,\min\,(s, t)$$

$$+ \frac{D}{\alpha}\,(-2 + 2\,e^{-\alpha t} + 2\,e^{-\alpha s} - e^{-\alpha|t-s|} - e^{-\alpha(t+s)}),$$

where, as is customary, we have set

$$D = \sigma^2/2\,\alpha^2.$$

If we now let α approach ∞ in such a way that D remains constant, we obtain

$$E\,Y_t \to E\,Y_0,$$

$$E\,(Y_s - E\,Y_s)\,(Y_t - E\,Y_t) \to \text{Var}\,(Y_0) + 2\,D\,\min\,(s, t),$$

that is, all finite-dimensional distributions of the Ornstein-Uhlenbeck process Y_t converge to the distributions of the Gaussian process

$$Y_t^{(0)} = Y_0 + \sqrt{2\,D}\,W_t.$$

But this is the Wiener process that starts at Y_0, multiplied by $\sqrt{2 D}$. In this sense, the Wiener process approximates the Ornstein-Uhlenbeck process. The sample functions of Y_t possess a derivative, namely, X_t. In the Ornstein-Uhlenbeck theory of Brownian motion, the particle therefore possesses a continuous velocity (but no acceleration), which ceases to exist when we shift to $Y_t^{(0)}$.

(8.3.1)   **Remark.** An analogous electrical problem leads formally to the same Langevin equation. Let X_t denote the current in an inductance-resistance circuit. Then,

$$L \dot{X}_t + R X_t = \sigma \xi_t,$$

where ξ_t is a rapidly fluctuating electromagnetic force generated by the thermal noise and again idealizable as a "white noise".

(8.3.2)   **Remark.** The Langevin equation for the position X_t of a Brownian particle in an external force field is

$$\ddot{X}_t + \beta \dot{X}_t = \sigma \xi_t + K (t, X_t).$$

By setting $V_t = \dot{X}_t$, we obtain from this equation the system

$$d \begin{pmatrix} V_t \\ X_t \end{pmatrix} = \begin{pmatrix} -\beta V_t + K (t, X_t) \\ V_t \end{pmatrix} dt + \begin{pmatrix} \sigma \\ 0 \end{pmatrix} dW_t.$$

We can find a closed solution in the case of a harmonic oscillator, for which $K (t, x) = -\nu^2 x$, so that the corresponding equation is linear (see Chandrasekhar [32], pp. 27-30).

8.4   The General Scalar Linear Equation

As preparation for the study of the vector-valued case, let us investigate first the case $d = 1$ (with m arbitrary) of the equation

$$(8.4.1) \qquad dX_t = (A (t) X_t + a (t)) dt + \sum_{i=1}^{m} (B_i (t) X_t + b_i (t)) dW_t^i, \quad X_{t_0} = c.$$

All the quantities in this equation (except $W_t \in R^m$) are scalar functions. Suppose that the coefficients A, a, B_i, and b_i are measurable and bounded on the interval $[t_0, T]$, so that there always exists a unique solution X_t, which we shall now determine explicitly.

(8.4.2)   **Theorem.** Equation (8.4.1) has the solution

$$X_t = \Phi_t \left(c + \int_{t_0}^{t} \Phi_s^{-1} \left(a (s) - \sum_{i=1}^{m} B_i (s) b_i (s) \right) ds + \sum_{i=1}^{m} \int_{t_0}^{t} \Phi_s^{-1} b_i (s) dW_s^i \right),$$

where

$$\Phi_t = \exp\left(\int_{t_0}^{t}\left(A(s) - \sum_{i=1}^{m} B_i(s)^2/2\right)ds + \sum_{i=1}^{m}\int_{t_0}^{t} B_i(s)\,dW_s^i\right)$$

is the solution of the homogeneous equation

$$d\Phi_t = A(t)\,\Phi_t\,dt + \sum_{i=1}^{m} B_i(t)\,\Phi_t\,dW_t^i$$

with initial value $\Phi_{t_0} = 1$.

Proof.   Let us use Itô's theorem to show that this process has the stochastic differential (8.4.1). If we now set $\Phi_t = \exp Y_t$ and

$$Z_t = c + \int_{t_0}^{t} e^{-Y_s}\left(a(s) - \sum_{i=1}^{m} B_i(s)\,b_i(s)\right)ds + \sum_{i=1}^{m}\int_{t_0}^{t} e^{-Y_s} b_i(s)\,dW_s^i,$$

we get

$$X_t = u(Y_t, Z_t)$$

where u is defined by

$$u(x, y) = e^x y.$$

Application of formula (5.3.9b) yields

$$dX_t = X_t\,dY_t + e^{Y_t}\,dZ_t + \frac{1}{2}\sum_{i=1}^{m}\text{tr}\begin{pmatrix} X_t & \Phi_t \\ \Phi_t & 0 \end{pmatrix}\begin{pmatrix} B_i^2\,\Phi_t^{-1}\,B_i\,b_i \\ \Phi_t^{-1}\,B_i\,b_i\,\Phi_t^{-2}\,b_i^2 \end{pmatrix}dt$$

$$= X_t\left(A(t) - \sum_{i=1}^{m} B_i(s)^2/2\right)dt + X_t\left(\sum_{i=1}^{m} B_i(t)\,dW_t^i\right)$$

$$+ \left(a(t) - \sum_{i=1}^{m} B_i(t)\,b_i(t)\right)dt + \sum_{i=1}^{m} b_i(t)\,dW_t^i$$

$$+ \sum_{i=1}^{m}\left(X_t\,B_i(t)^2/2 + B_i(t)\,b_i(t)\right)dt$$

$$= (A(t)\,X_t + a(t))\,dt + \sum_{i=1}^{m}(B_i(t)\,X_t + b_i(t))\,dW_t^i. \blacksquare$$

Example (6.3.11) is a special case of this result. If c has a normal distribution, X_t is a Gaussian process only when $B_1(t) = \dots = B_m(t) \equiv 0$ (i.e. the equation is linear in the narrow sense). Then, Theorem (8.4.2) reduces to Corollary (8.2.5).

We single out two special cases:

(8.4.3) **Corollary.** Suppose that $d = 1$. Then, the solution

a) of the *homogeneous* equation

$$\mathrm{d}X_t = A(t) X_t \,\mathrm{d}t + \sum_{i=1}^{m} B_i(t) X_t \,\mathrm{d}W_t^i, \quad X_{t_0} = c,$$

is

$$X_t = c \exp\left(\int_{t_0}^{t} \left(A(s) - \sum_{i=1}^{m} B_i(s)^2/2\right) \mathrm{d}s + \sum_{i=1}^{m} \int_{t_0}^{t} B_i(s) \,\mathrm{d}W_s^i\right),$$

b) of the *homogeneous autonomous* equation $(A(t) \equiv A, B_i(t) \equiv B_i)$

$$\mathrm{d}X_t = A X_t \,\mathrm{d}t + \sum_{i=1}^{m} B_i X_t \,\mathrm{d}W_t^i, \quad X_{t_0} = c,$$

is

$$X_t = c \exp\left(\left(A - \sum_{i=1}^{m} B_i^2/2\right)(t - t_0) + \sum_{i=1}^{m} B_i (W_t^i - W_{t_0}^i)\right).$$

We note that, by virtue of the law of large numbers for W_t, we have in the last case for arbitrary c

$$\mathrm{ac}\text{-}\lim_{t \to \infty} X_t = 0$$

provided

$$A < \sum_{i=1}^{m} B_i^2/2 \ .$$

In general, in the homogeneous case, X_t has, for all $t \in [t_0, T]$, the same sign as c. Let us now calculate the moments of X_t. For this we use

(8.4.4) **Lemma.** If X is $\Re(\alpha, \sigma^2)$-distributed, then, for every $p > 0$,

$$E(e^X)^p = e^{p\alpha + p^2 \sigma^2/2}.$$

Proof.

$$\frac{1}{\sqrt{2\pi\sigma^2}} \int_{-\infty}^{\infty} \exp(px - (x - \alpha)^2/2\sigma^2) \,\mathrm{d}x = \exp(p\alpha + p^2 \sigma^2/2). \quad \blacksquare$$

(8.4.5) **Theorem.** The solution X_t of the scalar linear stochastic differential equation (8.4.1) has, for all $t \in [t_0, T]$, a pth-order moment if and only if $E|c|^p < \infty$. In particular,

a)

$$m_t = E\, X_t = \varphi_t \left(E\, c + \int_{t_0}^{t} \varphi_s^{-1}\, a\,(s)\, \mathrm{d}s \right)$$

where

$$\varphi_t = \exp \left(\int_{t_0}^{t} A\,(s)\, \mathrm{d}s \right),$$

that is, m_t is the solution of the ordinary differential equation

$$\dot{m}_t = A\,(t)\, m_t + a\,(t), \quad m_{t_0} = E\, c.$$

b) $P\,(t) = E\, X_t^2$ is the unique solution of the ordinary linear differential equation

$$(8.4.6) \qquad \dot{P}\,(t) = \left(2\, A\,(t) + \sum_{i=1}^{m} B_i\,(t)^2 \right) P\,(t) + 2\, m_t \left(a\,(t) + \sum_{i=1}^{m} B_i\,(t)\, b_i\,(t) \right)$$
$$+ \sum_{i=1}^{m} b_i\,(t)^2$$

with initial condition

$$P\,(t_0) = E\, c^2,$$

where $m_t = E\, X_t$.

c)   In the homogeneous case $a\,(t) \equiv 0$, $b_i\,(t) \equiv 0$ (for $i = 1, \dots, m$), for the pth absolute moment of the solution

$$X_t = c \exp \left(\int_{t_0}^{t} \left(A\,(s) - \sum_{i=1}^{m} B_i\,(s)^2/2 \right) \mathrm{d}s + \sum_{i=1}^{m} \int_{t_0}^{t} B_i\,(s)\, \mathrm{d}W_s^i \right)$$

we have

$$E\, |X_t|^p = E\, |c|^p \exp \left(p \int_{t_0}^{t} \left(A\,(s) - \sum_{i=1}^{m} B_i\,(s)^2/2 \right) \mathrm{d}s + \frac{p^2}{2} \int_{t_0}^{t} \sum_{i=1}^{m} B_i\,(s)^2\, \mathrm{d}s \right).$$

(8.4.7)

Proof.   The homogeneous equation has the solution

$$X_t = c\, \Phi_t,$$

where c and Φ_t are statistically independent and Φ_t has moments of every order. Therefore,

$$E\, |X_t|^p = E\, |c|^p\, E\, \Phi_t^p$$

is finite if and only if $E |c|^p$ is finite. In the nonhomogeneous case, we add to the solution of the homogeneous equation only terms with finite moments of every order, so that it is a matter only of $E |c|^p$. We obtain the form of m_t immediately from (8.4.2) by using Lemma (8.4.4) with $p = 1$, and we obtain the differential equation for m_t either by differentiating this result or directly from the integral form of equation (8.4.1).

From example (5.4.9), we have

$$
dX_t^2 = 2 X_t (A (t) X_t + a (t)) dt + 2 X_t \sum_{i=1}^{m} (B_i (t) X_t + b_i (t)) dW_t^i
$$

$$
+ \sum_{i=1}^{m} (B_i (t) X_t + b_i (t))^2 dt.
$$

If we take the expectation on both sides of the integral form of this equation, we get

$$
P (t) = E c^2 + \int_{t_0}^{t} \left(2 A (s) P (s) + \sum_{i=1}^{m} B_i (s) P (s) \right) ds
$$

$$
+ \int_{t_0}^{t} 2 m_s \left(a (s) + \sum_{i=1}^{m} B_i (s) b_i (s) \right) ds + \sum_{i=1}^{m} \int_{t_0}^{t} b_i (s)^2 ds,
$$

and, after differentiation, equation (8.4.6).

Finally, (8.4.7) is a consequence of the independence of c and Φ_t and of Lemma (8.4.4). ∎

(8.4.8)   **Example.**   For the homogeneous autonomous equation

$$
dX_t = A X_t dt + \sum_{i=1}^{m} B_i X_t dW_t^i, \quad X_{t_0} = c,
$$

we have

$$
E |X_t|^p = E |c|^p \exp \left(p \left(A - \sum_{i=1}^{m} B_i^2 / 2 \right) (t - t_0) + \frac{p^2}{2} \sum_{i=1}^{m} B_i^2 (t - t_0) \right).
$$

Therefore,

$$
\lim_{t \to \infty} E |X_t|^p = \begin{cases} 0 \\ E |c|^p \\ + \infty \end{cases}
$$

if and only if

$$A \lesseqqgtr (1-p) \sum_{i=1}^{m} B_i^2 / 2 \, .$$

8.5 The General Vector Linear Equation

We now return to the general linear stochastic differential equation for an R^d-valued process X_t:

$$(8.5.1) \qquad \mathrm{d}X_t = (A\,(t)\,X_t + a\,(t))\,\mathrm{d}t + \sum_{i=1}^{m} (B_i\,(t)\,X_t + b_i\,(t))\,\mathrm{d}W_t^i,$$

where $A\,(t)$ and $B_i\,(t)$ are $d \times d$ matrices, $a\,(t)$ and $b_i\,(t)$ are R^d-valued functions, and $W_t = (W_t^1, \ldots, W_t^m)'$ is an m-dimensional Wiener process. By Theorem (8.1.5), there exists, for every initial value c that is independent of $W_t - W_{t_0}$ for $t \in [t_0, T]$, a unique solution of (8.5.1) on the interval $[t_0, T]$, provided the coefficients A, a, B_i, and b_i are measurable bounded functions on that interval, as we shall always assume.

We now model the general solution of (8.5.1) after the scalar case $d = 1$, so that it will include the case treated in section 8.2 as a special case.

(8.5.2) **Theorem.** The linear stochastic differential equation (8.5.1) with initial value $X_{t_0} = c$ has on $[t_0, T]$ the solution

$$(8.5.3) \qquad X_t = \Phi_t \left(c + \int_{t_0}^{t} \Phi_s^{-1}\,\mathrm{d}Y_s \right).$$

Here,

$$\mathrm{d}Y_t = \left(a\,(t) - \sum_{i=1}^{m} B_i\,(t)\,b_i\,(t) \right)\mathrm{d}t + \sum_{i=1}^{m} b_i\,(t)\,\mathrm{d}W_t^i,$$

and the $d \times d$ matrix Φ_t is the fundamental matrix of the corresponding homogeneous equation, that is, the solution of the homogeneous stochastic differential equation

$$\mathrm{d}\Phi_t = A\,(t)\,\Phi_t\,\mathrm{d}t + \sum_{i=1}^{m} B_i\,(t)\,\Phi_t\,\mathrm{d}W_t^i$$

with initial value

$$\Phi_{t_0} = I \, .$$

Proof. Since a unique solution is known to exist, it will be sufficient to verify directly that the given X_t satisfies equation (8.5.1). To do this, we set

$$Z_t = c + \int_{t_0}^{t} \Phi_s^{-1} \, dY_s,$$

$$dZ_t = \Phi_t^{-1} \, dY_t,$$

and use Itô's theorem to calculate the stochastic differential of

$$X_t = \Phi_t Z_t.$$

Application of example (5.4.1) to each component of X_t yields

$$dX_t = \Phi_t \, dZ_t + (d\Phi_t) \, Z_t + \left(\sum_{i=1}^{m} B_i(t) \, \Phi_t \, \Phi_t^{-1} \, b_i(t) \right) dt$$

$$= dY_t + A(t) \, X_t \, dt + \sum_{i=1}^{m} B_i(t) \, X_t \, dW_t^i + \sum_{i=1}^{m} B_i(t) \, b_i(t) \, dt$$

$$= (a(t) + A(t) \, X_t) \, dt + \sum_{i=1}^{m} (B_i(t) \, X_t + b_i(t)) \, dW_t^i.$$

The initial value is

$$X_{t_0} = \Phi_{t_0} Z_{t_0} = I \, c = c. \quad \blacksquare$$

(8.5.4) **Remark.** In complete analogy with ordinary differential equations, the general solution (8.5.3) can be represented as a sum as follows:

$$X_t = \Phi_t \, c + \Phi_t \int_{t_0}^{t} \Phi_s^{-1} \, dY_s.$$

Here, the first term on the right is the general solution of the corresponding homogeneous equation (which here in contrast with Theorem (8.2.2) is in general also a stochastic process), and the second term is the particular solution of the nonhomogeneous equation corresponding to the initial value $X_{t_0} = 0$. For $d = 1$, we have Φ_t given explicitly in Theorem (8.4.2). For $B_i(t) \equiv 0$, Theorem (8.5.2) reduces to Theorem (8.2.2).

Let us look again at the first-order ordinary differential equations that the first two moments of the solution must satisfy.

(8.5.5) **Theorem.** For the solution (8.5.3) of the linear stochastic differential equation (8.5.1), we have under the assumption $E \, |c|^2 < \infty$,

a) $E \, X_t = m_t$ is the unique solution of the equation

$$\dot{m}_t = A(t) \, m_t + a(t), \quad m_{t_0} = E \, c.$$

b) $E \, X_t X_t' = P(t)$ is the unique nonnegative-definite symmetric solution of the equation

$$\dot{P}(t) = A(t) P(t) + P(t) A(t)' + a(t) m_t' + m_t a(t)'$$

(8.5.6)

$$+ \sum_{i=1}^{m} (B_i(t) P(t) B_i(t)' + B_i(t) m_t b_i(t)' + b_i(t) m_t' B_i(t)' + b_i(t) b_i(t)')$$

with initial value

$$P(t_0) = E c c'.$$

Proof. Part a) follows when we take the expectation on both sides of the integral form of (8.5.1). Part b) follows in the same way from

$$dX_t X_t' = X_t dX_t' + (dX_t) X_t' + \sum_{i=1}^{m} (B_i(t) X_t + b_i(t)) (X_t' B_i(t)' + b_i(t)') dt$$

$$= \Big(X_t X_t' A(t) + X_t a(t)' + A(t) X_t X_t' + a(t) X_t'$$

$$+ \sum_{i=1}^{m} (B_i(t) X_t X_t' B_i(t)' + B_i(t) X_t b_i(t)'$$

$$+ b_i(t) X_t' B_i(t)' + b_i(t) b_i(t)') \Big) dt$$

$$+ \sum_{i=1}^{m} (X_t X_t' B_i(t)' + X_t b_i(t)' + B_i(t) X_t X_t' + b_i(t) X_t') dW_t^i.$$

This formula is obtained from Itô's theorem. Both equations have unique solutions on the interval $[t_0, T]$ since the right-hand members satisfy the boundedness and Lipschitz conditions. Since

$$P(t) = P(t)',$$

(8.5.6) represents a system of $d(d+1)/2$ linear equations. The solution $P(t)$, being the covariance matrix of X_t, is of course nonnegative-definite. ∎

(8.5.7) **Remark.** The function $m_t = E X_t$ is independent of the fluctuational part (that is, independent of the B_i and the b_i) of equation (8.5.1).

(8.5.8) **Remark.** We have

$$E |X_t|^2 = \operatorname{tr} P(t) = \sum_{i=1}^{d} P_{ii}(t).$$

However, the differential equation for $E |X_t|^2$, which follows from (8.5.6), contains in general in its right-hand member all other elements of the matrix $P(t)$.

(8.5.9) **Remark.** Even if the homogeneous equation

$$dX_t = A(t) X_t \, dt + \sum_{i=1}^{m} B_i(t) X_t \, dW_t^i$$

corresponding to (8.5.1) is autonomous, that is, if $A(t) \equiv A$ and $B_i(t) \equiv B_i$ are independent of t, the corresponding fundamental solution Φ_t nonetheless cannot in general be given explicitly. Only when, for example, the matrices $A, B_1, \ldots, B_m$ commute, that is, when

$$A B_i = B_i A, \quad B_i B_j = B_j B_i, \quad \text{all } i, j,$$

does the fundamental matrix assume the form

$$\Phi_t = \exp\left(\left(A - \sum_{i=1}^{m} B_i^2/2\right)(t - t_0) + \sum_{i=1}^{m} B_i (W_t^i - W_{t_0}^i)\right)$$

To show this, we set

$$dY_t = \left(A - \sum_{i=1}^{m} B_i^2/2\right) dt + \sum_{i=1}^{m} B_i \, dW_t^i, \quad Y_{t_0} = 0,$$

and calculate the stochastic differential of

$$\Phi_t = \exp(Y_t).$$

By virtue of the commutativity of the participating matrices, we have

$$d\Phi_t = \exp(Y_t) \, dY_t + \frac{1}{2} \exp(Y_t) (dY_t)^2$$

$$= \Phi_t \, dY_t + \frac{1}{2} \Phi_t \left(\sum_{i=1}^{m} B_i^2\right) dt$$

$$= A \Phi_t \, dt + \sum_{i=1}^{m} B_i \Phi_t \, dW_t^i,$$

that is, Φ_t satisfies the homogeneous equation.

Chapter 9

The Solutions of Stochastic Differential Equations as Markov and Diffusion Processes

9.1 Introduction

In the preceding three chapters, we have constructed and examined in a sort of "stochastic analysis" the solutions of stochastic differential equations. This puts us in a position to calculate explicitly, for given sample functions of the initial value and of the Wiener process, an arbitrarily accurate solution trajectory, for example, with the aid of the iteration procedure used in the proof of Theorem (6.2.2).

On the other hand, the solution X_t is a stochastic process on the interval $[t_0, T]$ and, as such, it can be regarded as a set of compatible finite-dimensional distributions

$$P[X_{t_1} \in B_1, \dots, X_{t_n} \in B_n] = P_{t_1, \dots, t_n}(B_1, \dots, B_n).$$

In accordance with Theorem (2.2.5), for the important class of Markov processes all these distributions can be obtained from the initial probability

$$P[X_{t_0} \in B] = P_{t_0}(B)$$

and the transition probability

$$P(X_t \in B \mid X_s = x) = P(s, x, t, B), \quad t_0 \leqq s \leqq t \leqq T.$$

Specifically,

$$P[X_{t_1} \in B_1, \dots, X_{t_n} \in B_n] = \int_{R^d} \int_{B_1} \cdots \int_{B_{n-1}} P(t_{n-1}, x_{n-1}, t_n, B_n) \cdot$$

$$\cdot P(t_{n-2}, x_{n-2}, t_{n-2}, dx_{n-1}) \dots P(t_1, x_1, t_2, dx_2) P(t_0, x_0, t_1, dx_1) P_{t_0}(dx_0),$$

$$t_0 < t_1 < \dots < t_n \leqq T, \quad B_i \in \mathcal{B}^d.$$

Stochastic differential equations owe their significance and expanding study not least to the fact that, as we shall show, their solutions are Markov processes. Therefore, we have for them the powerful analytical tools developed for Markov processes at our disposal. The keystone of the Markov property of the solution processes is the fact that a white noise ξ_t in the form

$$\dot{X}_t = f(t, X_t) + G(t, X_t)\,\xi_t$$

of the stochastic differential equation

$$dX_t = f(t, X_t)\,dt + G(t, X_t)\,dW_t$$

is a process with independent values at every point.

Furthermore, in many cases X_t is in fact a diffusion process whose drift vector and diffusion matrix can be read in the simplest conceivable manner from the equation (see section 9.3).

9.2   The Solutions as Markov Processes

Consider the stochastic differential equation

$$(9.2.1) \qquad dX_t = f(t, X_t)\,dt + G(t, X_t)\,dW_t, \quad X_{t_0} = c,$$

on the interval $[t_0, T]$. Here, X_t and f assume values in R^d, G is $(d \times m$ matrix)-valued, and W_t is an R^m-valued Wiener process. The initial value c is an arbitrary random variable independent of $W_t - W_{t_0}$ for $t \geq t_0$.

Together with (9.2.1), let us consider the same equation, but now on the interval $[s, T]$, for $t_0 \leq s \leq T$ and with the fixed initial value $X_s = x \in R^d$, hence the equivalent integral equation

$$(9.2.2) \qquad X_t = x + \int\limits_s^t f(u, X_u)\,du + \int\limits_s^t G(u, X_u)\,dW_u, \quad t_0 \leq s \leq t \leq T.$$

(9.2.3)   **Theorem.** If equation (9.2.1) satisfies the conditions of the existence-and-uniqueness theorem (6.2.2), the solution X_t of the equation for arbitrary initial values is a *Markov process* on the interval $[t_0, T]$ whose initial probability distribution at the instant t_0 is the distribution of c and whose transition probabilities are given by

$$P(s, x, t, B) = P(X_t \in B \mid X_s = x) = P[X_t(s, x) \in B]$$

where $X_t(s, x)$ is the (existent and unique) solution of equation (9.2.2).

Proof. Let $(\Omega, \mathfrak{A}, P)$ denote the basic probability space on which c and W_t, for $t \geq 0$, are defined. As usual, let $\mathfrak{F}_t \subset \mathfrak{A}$ denote the sigma-algebra generated by c and W_s for $s \leq t$, which is independent of the sigma-algebra $\mathfrak{W}_t^+$ generated by $W_s - W_t$ for $s \geq t$. We need also to consider the sigma-algebras

$$\mathfrak{A}_t = \mathfrak{A}\left([t_0, t]\right) = \mathfrak{A}\left(X_s, t_0 \leqq s \leqq t\right), \quad t_0 \leqq t \leqq T,$$

which contain the "history" of the process X_t up to the instant t. We need to prove the Markov property for X_t (see definition (2.1.1)): For $t_0 \leqq s \leqq t \leqq T$ and all $B \in \mathfrak{B}^d$, we have

(9.2.4) $P\left(X_t \in B \mid \mathfrak{A}_s\right) = P\left(X_t \in B \mid X_s\right),$

almost certainly $[P]$. Now, X_t is $\mathfrak{F}_t$-measurable (that is, nonanticipating), so that

$$\mathfrak{A}_t \subset \mathfrak{F}_t.$$

The validity of (9.2.4) therefore follows from the stronger equation

(9.2.5) $P\left(X_t \in B \mid \mathfrak{F}_s\right) = P\left(X_t \in B \mid X_s\right)$

by virtue of (1.7.1).

Furthermore, instead of (9.2.5), it will be sufficient to prove the following: For every scalar bounded measurable function $h\left(x, \omega\right)$ defined on $R^d \times \Omega$ for which $h\left(x, \cdot\right)$ is, for every fixed x, a random variable independent of $\mathfrak{F}_s$, we have

(9.2.6) $E\left(h\left(X_s, \omega\right) \mid \mathfrak{F}_s\right) = E\left(h\left(X_s, \omega\right) \mid X_s\right) = H\left(X_s\right)$

with $H\left(x\right) = E\, h\left(x, \omega\right)$. This is true because, if we choose

$$h\left(x, \omega\right) = I_B\left(X_t\left(s, x, \omega\right)\right),$$

where $X_t\left(s, x, \omega\right) = X_t\left(s, x\right)$ is the solution of equation (9.2.2) and I_B is the indicator function of the set B, then $h\left(x, \cdot\right)$ is independent of $\mathfrak{F}_s$ since, by virtue of the constant initial point at the instant s, the function $X_t\left(s, x\right)$ is $\mathfrak{W}_s^+$-measurable. By virtue of remark (6.1.7), we have

$$X_t = X_t\left(t_0, c\right) = X_t\left(s, X_s\left(t_0, c\right)\right) = X_t\left(s, X_s\right)$$

so that

$$h\left(X_s, \omega\right) = I_B\left(X_t\right).$$

Therefore, in this case, equation (9.2.6) yields

$$P\left(X_t \in B \mid \mathfrak{F}_s\right) = P\left(X_t \in B \mid X_s\right) = P\left[X_t\left(s, x\right) \in B\right]_{x = X_s},$$

from which, by virtue of (9.2.5), not only the Markov property (9.2.4) but also the asserted form

$$P\left(s, x, t, B\right) = P\left[X_t\left(s, x\right) \in B\right]$$

of the transition probabilities follows.

Thus, it remains to prove (9.2.6). We shall do this for the set of functions of the form

$$h\left(x, \omega\right) = \sum_{i=1}^{n} Y_i\left(x\right) Z_i\left(\omega\right), \quad Z_i \text{ independent of } \mathfrak{F}_s,$$

which is dense in the set of bounded measurable functions h under consideration. For functions of this form, we have

$$E\left(h\left(X_s, \omega\right)|\mathfrak{F}_s\right) = \sum_{i=1}^{n} Y_i\left(X_s\right) E\left(Z_i\right) = H\left(X_s\right)$$

with $H\left(x\right) = E\,h\left(x, \omega\right)$. Since X_s is of course $\mathfrak{A}\left(X_s\right)$-measurable and Z_i is independent of X_s, we have in addition

$$\sum_{i=1}^{n} Y_i\left(X_s\right) E\left(Z_i\right) = E\left(h\left(X_s, \omega\right)|X_s\right),$$

which proves (9.2.6).

Since $X_{t_0} = c$, we have, for the initial probability of X_t,

$$P_{t_0}\left(B\right) = P\left[X_{t_0} \in B\right] = P\left[c \in B\right]. \quad \blacksquare$$

(9.2.7) **Remark.** Whereas the initial probability of the process X_t is identical to the distribution of c, the transition probabilities do not depend on c but are completely determined by the coefficients f and G, which can be read from equation (9.2.1), and hence by the "system". Furthermore, the transition probabilities of X_t acquire, by virtue of the formula

$$P\left(X_t \in B|X_s = x\right) = P\left[X_t\left(s, x\right) \in B\right]$$

not only in a heuristic but also in a strictly mathematical way, the following significance: The (conditional) probability of the event $[X_t \in B]$ under the condition $X_s = x$ is equal to the absolute probability of the event $[X_t\left(s, x\right) \in B]$, where the process $X_t\left(s, x\right)$ begins at the instant s with probability 1 at x.

Theorem (9.2.3) can be successfully used for practical calculation of the transition probabilities only when we know explicitly the solution $X_t\left(s, x\right)$ of equation (9.2.2) or at least its distribution. Better methods for calculating the function $P\left(s, x, t, B\right)$ or the density $p\left(s, x, t, y\right)$ will be presented in section 9.4.

In accordance with definition (2.2.9), we call a Markov process **homogeneous** if its transition probabilities are stationary, that is, if the condition

$$P\left(s+u, x, t+u, B\right) = P\left(s, x, t, B\right), \quad 0 \leq u \leq T-t,$$

is satisfied identically. In this case, the function $P\left(s, x, t, B\right) = P\left(t_0, x, t_0 + (t-s), B\right) = P\left(t-s, x, B\right)$ is therefore a function only of $x \in R^d$, $t-s \in [0, T-t_0]$, and $B \in \mathfrak{B}^d$.

(9.2.8) **Theorem.** Suppose that the conditions of the existence-and-uniqueness theorem (6.2.2) are satisfied for equation (9.2.1). If the coefficients $f\left(t, x\right) \equiv f\left(x\right)$ and $G\left(t, x\right) \equiv G\left(x\right)$ are independent of t on the interval $[t_0, T]$, then the solution X_t is, for arbitrary initial values c, a *homogeneous Markov process* with the (stationary) transition probabilities

$$P(X_t \in B | X_{t_0} = x) = P(t - t_0, x, B) = P[X_t(t_0, x) \in B],$$

where $X_t(t_0, x)$ is the solution of equation (9.2.1) with initial value $X_{t_0} = x$. In particular, the solution of an *autonomous* equation

$$dX_t = f(X_t)\,dt + G(X_t)\,dW_t, \quad t \geqq t_0 ,$$

is a homogeneous Markov process defined for all $t \geqq t_0$.

The assertions of this theorem are intuitively so clear that we refrain from giving a proof.

(9.2.12) **Example.** In accordance with Theorem (8.2.2), the linear (in the narrow sense) stochastic differential equation

$$dX_t' = (A(t) X_t + a(t))\,dt + B(t)\,dW_t, \quad X_{t_0} = c, \quad t_0 \leqq t \leqq T,$$

has, corresponding to the initial value x at the instant s, the solution

$$X_t(s, x) = \Phi(t, s)\left(x + \int_s^t \Phi(u, s)^{-1} a(u)\,du + \int_s^t \Phi(u, s)^{-1} B(u)\,dW_u\right),$$

where $\Phi(t, s)$ is the solution of the homogeneous matrix equation

$$\frac{d}{dt}\Phi(t, s) = A(t)\,\Phi(t, s), \quad \Phi(s, s) = I.$$

From Theorems (8.2.6) and (8.2.10), the transition probability $P(s, x, t, \cdot)$ of X_t is a d-dimensional normal distribution

$$P(s, x, t, \cdot) = \mathfrak{N}(m_t(s, x), K_t(s, x)),$$

with expectation vector

$$\int_{R^d} y\, P(s, x, t, dy) = E X_t(s, x) = m_t(s, x)$$

$$= \Phi(t, s)\left(x + \int_s^t \Phi(u, s)^{-1} a(u)\,du\right)$$

and $d \times d$ covariance matrix

$$\int_{R^d} (y - m_t(s, x))(y - m_t(s, x))'\, P(s, x, t, dy) =$$

$$= E(X_t(s, x) - m_t(s, x))(X_t(s, x) - m_t(s, x))'$$

$$= K_t(s, x)$$

$$= \Phi(t, s) \int_s^t \Phi(u, s)^{-1} B(u) B(u)' (\Phi(u, s)^{-1})' \, du \, \Phi(t, s)'.$$

In accordance with Theorem (8.2.10), for Gaussian or constant c the process X_t is itself a Gaussian process, frequently known as the **Gauss-Markov process.**

In the autonomous case $A(t) \equiv A$, $a(t) \equiv a$, $B(t) \equiv B$, the transition probability of the now homogeneous Markov process X_t is, by virtue of the relationship

$$\Phi(t, s) = e^{A(t-s)}$$

specialized to

$$P(s, x, s+t, \cdot) = P(t, x, \cdot) = \mathfrak{N}(m_{s+t}(s, x), K_{s+t}(s, x))$$

with the functions

$$m_{s+t}(s, x) = e^{At} \left(x + \left(\int_0^t e^{-Au} \, du \right) a \right)$$

and

$$K_{s+t}(s, x) = \int_0^t e^{A(t-u)} B B' e^{A'(t-u)} \, du$$

depending only on t and x. In the case in which A is nonsingular and A and B commute (that is, $AB = BA$), these expressions are simplified yet further:

$$m_{s+t}(s, x) = e^{At}(x + A^{-1} a) - A^{-1} a$$

and

$$K_{s+t}(s, x) = B(A + A')^{-1} (e^{(A+A')t} - I) B'.$$

(9.2.13) **Example.** In accordance with Corollary (8.4.3), the scalar linear homogeneous stochastic differential equation ($d = 1$, arbitrary m)

$$dX_t = A(t) X_t \, dt + \sum_{i=1}^m B_i(t) X_t \, dW_t^i, \quad X_{t_0} = c, \quad t_0 \le t \le T,$$

with initial value x at the instant s has the solution

$$X_t(s, x) = x \exp \left(\int_s^t \left(A(u) - \sum_{i=1}^m B_i(u)^2/2 \right) du + \sum_{i=1}^m \int_s^t B_i(u) \, dW_u^i \right).$$

Since $X_t(s, 0) \equiv 0$, we have $P(s, 0, t, \cdot) = \delta_0$ and, since $X_t(s, -x) \equiv -X_t(s, x)$, we have $P(s, -x, t, B) = P(s, x, t, -B)$. Therefore, we can confine ourselves to positive x. Then, $X_t(s, x)$ is also positive on $[s, T]$ and we have, for $y > 0$,

$$P(s, x, t, (0, y]) = P[X_t(s, x) \le y]$$

$$= P\left[\sum_{i=1}^{m} \int_s^t B_i(u)\, dW_u^i \le \log\left(\frac{y}{x}\right) - \int_s^t \left(A(u) - \sum_{i=1}^{m} B_i(u)^2/2\right) du\right]$$

$$= (\sqrt{2\pi}\, \sigma)^{-1} \int_{-\infty}^{z} e^{-u^2/2\sigma^2}\, du$$

where

$$z = \log(y/x) - \int_s^t \left(A(u) - \sum_{i=1}^{m} B_i(u)^2/2\right) du$$

and

$$\sigma^2 = \sum_{i=1}^{m} \int_s^t B_i(u)^2\, du.$$

In the case of an autonomous equation $(A(t) \equiv A, B_i(t) \equiv B_i)$, the transition probability is stationary with the parameters.

$$z = \log(y/x) - \left(A - \sum_{i=1}^{m} B_i^2/2\right)(t-s)$$

and

$$\sigma^2 = \left(\sum_{i=1}^{m} B_i^2\right)(t-s).$$

The moments of $P(s, x, t, \cdot)$ can be obtained from Theorem (8.4.5c).

(9.2.14) **Remark.** When is the solution X_t of (9.2.1) a *stationary* Markov process? In accordance with remark (2.2.11), X_t must always be homogeneous, which is the case for the solution of an autonomous equation. For the existence of a stationary distribution P, that is, a distribution with the property

$$P(B) = \int_{R^d} P(t, x, B)\, dP(x), \quad B \in \mathfrak{B}^d, \quad t \ge 0,$$

there exist analytical conditions (see Prohorov-Razanov [15], pp. 272-274, Khas'minskiy [65], p. 119ff, and Itô and Nisio [43]). In many cases, the density $p(x)$ of the stationary distribution can be obtained from the stationary form of the forward equation (see section 9.4)

$$\sum_{i=1}^{d} \frac{\partial}{\partial x_i} (f_i(x) \, p(x)) - \frac{1}{2} \sum_{i=1}^{d} \sum_{j=1}^{d} \frac{\partial^2}{\partial x_i \, \partial x_j} ((G(x) \, G(x)')_{ij} \, p(x)) = 0$$

See also Theorem (8.2.12).

9.3   The Solutions as Diffusion Processes

According to definition (2.5.1), diffusion processes are Markov processes with continuous sample functions whose transition probabilities $P(s, x, t, B)$ have certain infinitesimal properties as $t \to s$.

The solutions of stochastic differential equations are Markov processes with continuous sample functions. When are these solutions diffusion processes? We would suppose that the coefficients f and G determining the transition probabilities need to satisfy additional conditions. Then, how are the drift and diffusion coefficients determined from the equation?

For simple cases, we can verify directly that the solution is a diffusion process. For example, if $f(t, x) \equiv f_0$ and $G(t, x) \equiv G_0$, then the solution of

$$dX_t = f_0 \, dt + G_0 \, dW_t, \quad X_{t_0} = c,$$

is the process

$$X_t = c + f_0 (t - t_0) + G_0 (W_t - W_{t_0}),$$

for whose transition probabilities, we have, in accordance with example (9.2.12)

$$P(s, x, t, \cdot) = \mathfrak{N}(x + f_0 (t - s), G_0 G_0' (t - s)).$$

It follows that

$$E_{s,x}(X_t - x) = f_0 (t - s),$$
$$E_{s,x}(X_t - x)(X_t - x)' = G_0 G_0' (t - s) + f_0 f_0' (t - s)^2,$$

where we have used the notation introduced in section 2.5:

$$E_{s,x} g(X_t) = \int_{R^d} g(y) \, P(s, x, t, dy)$$

In accordance with remark (2.5.2), X_t is a d-dimensional diffusion process with drift vector f_0 and diffusion matrix $G_0 G_0'$.

More generally, we have

(9.3.1)   **Theorem.** The conditions of the existence-and-uniqueness theorem (6.2.2) are satisfied for the stochastic differential equation

$$dX_t = f(t, X_t) \, dt + G(t, X_t) \, dW_t, \quad X_{t_0} = c, \quad t_0 \leq t \leq T,$$

where X_t and $f(t, x)$ belong to R^d, W_t belongs to R^m, and $G(t, x)$ is a $d \times m$ matrix. If in addition, the functions f and G are *continuous with respect to t*, the solution X_t is a d-dimensional diffusion process on $[t_0, T]$ with *drift vector* $f(t, x)$ and *diffusion matrix*

$$B(t, x) = G(t, x) \, G(t, x)'.$$

The limit relationships in definition (2.5.1) hold *uniformly* in $s \in [t_0, T)$, where $T < \infty$. In particular, the solution of an *autonomous* stochastic differential equation is always a *homogeneous diffusion process* on $[t_0, \infty)$.

Proof. From remark (7.1.5), we have

$$E_{s,x} |X_t - X_s|^4 = E |X_t(s, x) - x|^4 \leq C_1 (t - s)^2, \quad t_0 \leq s \leq t \leq T,$$

so that

$$\lim_{t \downarrow s} (t - s)^{-1} \int_{R^d} |y - x|^4 \, P(s, x, t, dy) = 0.$$

To prove that X_t is a diffusion process with given drift and diffusion coefficients, it is therefore sufficient, by virtue of remark (2.5.2), to show that

(9.3.2) $\qquad E(X_t(s, x) - x) = f(s, x)(t - s) + o(t - s)$

and

(9.3.3) $\qquad E(X_t(s, x) - x)(X_t(s, x) - x)' = G(s, x) \, G(s, x)'(t - s) + o(t - s).$

In particular, the asserted uniformity is obtained from the fact that $o(t - s)$ is independent of s.

We begin with equation (9.3.2) and write the left-hand side in the following form:

$$E X_t(s, x) - x = \int_s^t E f(u, X_u(s, x)) \, du$$

$$= \int_s^t f(u, x) \, du + \int_s^t (E f(u, X_u(s, x)) - f(u, x)) \, du.$$

Using the Schwarz inequality, the Lipschitz condition, and inequality (7.1.4), we obtain for $c = x$ and $n = 1$

$$\left| \int_s^t E(f(u, X_u(s, x)) - f(u, x)) \, du \right| \leq \int_s^t E |f(u, X_u(s, x)) - f(u, x)| \, du$$

$$\leq (t - s)^{1/2} \left(\int_s^t E |f(u, X_u(s, x)) - f(u, x)|^2 \, du \right)^{1/2}$$

$$\leqq 0\,((t-s)^{1/2})\left(\int\limits_{s}^{t} E\,|X_u\,(s,x)-x|^2\,du\right)^{1/2}$$

$$= (t-s)^{3/2}\,0\,(1).$$

The continuity of $f(\cdot,x)$ now implies that

$$\int\limits_{s}^{t} f(u,x)\,du = f(s,x)\,(t-s) + \int\limits_{s}^{t}(f(u,x)-f(s,x))\,du$$

$$= f(s,x)\,(t-s) + o\,(t-s),$$

so that

$$E\,X_t\,(s,x)-x = f(s,x)\,(t-s) + o\,(t-s),$$

which is equation (9.3.2).

Equation (9.3.3) is proven in a completely analogous manner. ∎

(9.3.4) **Example.** The (existing and unique) solution of the linear stochastic differential equation

$$dX_t = (a\,(t)+A\,(t)\,X_t)\,dt + \sum_{i=1}^{m}(B_i\,(t)\,X_t+b_i\,(t))\,dW_t^i,$$

for $t_0 \leqq t \leqq T$, is certainly a diffusion process if the functions $a\,(t)$, $A\,(t)$, $B_i\,(t)$, and $b_i\,(t)$ are continuous on the interval $[t_0, T]$. In particular, the solution of the autonomous linear equation is always a homogeneous diffusion process (see example (9.2.12)). The drift vector of X_t is

$$f(t,x) = a\,(t)+A\,(t)\,x$$

and the diffusion matrix is

$$B\,(t,x) = \sum_{i=1}^{m}(B_i\,(t)\,x+b_i\,(t))\,(x'\,B_i\,(t)'+b_i\,(t)')$$

$$= \sum_{i=1}^{m}(B_i\,x\,x'\,B_i' + B_i\,x\,b_i' + b_i\,x'\,B_i' + b_i\,b_i').$$

(9.3.5) **Remark.** We shall now discuss the opposite question as to when a given diffusion process is the solution of a stochastic differential equation. In other words, if X_t is a d-dimensional diffusion process on the interval $[t_0, T]$, do there exist a Wiener process W_t such that X_{t_0} and $W_t - W_{t_0}$ are statistically independent and functions f and G such that the sample functions of X_t can be obtained by means of the equation

$$dX_t = f\,(t,X_t)\,dt + G\,(t,X_t)\,dW_t$$

from the sample functions of W_t? After all, in accordance with Remark (6.1.4), this equation represents a transformation that maps $W.(\omega)$ and $X_{t_0}(\omega)$ into $X.(\omega)$. Sufficient conditions for this can be found, for example, in Prohorov and Rozanov [15], pp. 261-262 or Gikhman and Skorokhod [36], p. 70.

If, for a given diffusion process X_t with drift vector $f(t, x)$ and diffusion matrix $B(t, x)$, we wish to find a stochastic differential equation whose solution coincides with X_t only in the initial distribution P_{t_0} and the transition probabilities $P(s, x, t, B)$ (and hence in all finite-dimensional distributions), in other words, if we wish to reproduce not the given realizations of X_t but only their distributions, we proceed as follows: We choose a probability space $(\Omega, \mathfrak{A}, P)$ on which an m-dimensional Wiener process W_t and a random variable c independent of $W_t - W_{t_0}$ for $t \geqq t_0$ with distribution P_{t_0} can be defined and we consider the stochastic differential equation

(9.3.6) $dY_t = f(t, Y_t)\,dt + G(t, Y_t)\,dW_t, \quad Y_{t_0} = c, \quad t_0 \leqq t \leqq T.$

Here, $G(t, x)$ is a $d \times m$ matrix with the property

(9.3.7) $B(t, x) = G(t, x)\,G(t, x)'.$

Now, there are various possibilities for decomposing a given symmetric nonnegative-definite $d \times d$ matrix $B(t, x)$ in the form (9.3.7), so that the coefficient G in equation (9.3.6) is not uniquely determined. If we represent B in the form

$$B = U \Lambda U'$$

(where Λ is the diagonal matrix of the eigenvalues $\lambda_i \geqq 0$ (arranged in increasing order) and U is the orthogonal $d \times d$ matrix of the column eigenvectors u_i of B [see remark (5.2.4)]), then the choice $d = m$ and

$$G = U \Lambda^{1/2} U' = B^{1/2}$$

yields again a symmetric nonnegative matrix G, while

$$G = U \Lambda^{1/2} = (\sqrt{\lambda_1}\, u_1, \dots, \sqrt{\lambda_d}\, u_d)$$

has the advantage of pointing the column vectors of G in the direction of the eigenvectors of B. If k of the λ_i are identically equal to 0, we can go on to the $d \times m$ matrix

$$G = (\sqrt{\lambda_{k+1}}\, u_{k+1}, \dots, \sqrt{\lambda_d}\, u_d), \quad m = d - k.$$

This lack of uniqueness is not important, however, if the given diffusion process is uniquely determined by its parameters $f(t, x)$ and $B(t, x)$ (in the sense of unique determination of the transition probabilities $P(s, x, t, B)$ by f and B). For this nontrivial property (we have obtained f and B from the first and second moments of $P(s, x, t, B)$ only!) we have given a sufficient condition in remark (2.6.5). If the given process is uniquely determined by f and B, then all equations of the form (9.3.6) in which G is chosen on the basis of (9.3.7) and which satisfy the assumptions of Theorem (9.3.1) lead to the *same* diffusion process. In particular, for homogeneous diffusion processes, an autonomous equation can always be found as a dynamic model.

Summing up, we may say that *the solutions of stochastic differential equations and diffusion processes represent essentially the same classes of processes* despite their completely different definitions.

9.4 Transition Probabilities

In Theorem (9.2.3), we pointed out that the transition probabilities $P(s, x, t, B)$ of the solution X_t of the equation

$$(9.4.1) \qquad dX_t = f(t, X_t)\, dt + G(t, X_t)\, dW_t, \quad X_{t_0} = c, \quad t_0 \leqq t \leqq T,$$

are equal to the ordinary probability distributions of the solution $X_t(s, x)$ that begins at x at the instant s:

$$P(s, x, t, B) = P[X_t(s, x) \in B], \quad t_0 \leqq s \leqq t \leqq T, \quad x \in R^d, \quad B \in \mathfrak{B}^d.$$

In the notation of Chapter 2,

$$E_{s, x}\, g(t, X_t) = \int_{R^d} g(t, y)\, P(s, x, t, dy) .$$

Therefore,

$$E_{s, x}\, g(t, X_t) = E\, g(t, X_t(s, x)),$$

and, in particular,

$$E_{s, x}\, g(X_t) = E\, g(X_t(s, x)).$$

A method of determining the transition probabilities $P(s, x, t, B)$ of X_t without having to solve equation (9.4.1) is the subject of the present section. Equation (9.4.1) represents the law of development for the state X_t of the stochastic dynamic system under consideration. If we turn to the law of development for the functions $P(s, x, t, B)$, what we are doing is shifting from a stochastic differential equation to a second-order partial differential equation.

From here on, we shall assume that the conditions of Theorem (9.3.1) are satisfied, so that X_t is a diffusion process. The differential operator (2.6.1) corresponding to the process X_t is

$$(9.4.2) \qquad \mathfrak{D} = \sum_{i=1}^{d} f_i(s, x) \frac{\partial}{\partial x_i} + \frac{1}{2} \sum_{i=1}^{d} \sum_{j=1}^{d} b_{ij}(s, x) \frac{\partial^2}{\partial x_i\, \partial x_j},$$

$$B(s, x) = (b_{ij}(s, x)) = G(s, x)\, G(s, x)', \quad f = (f_1, \dots, f_d)'$$

where the derivatives are evaluated at the point (s, x). The infinitesimal operator A, which uniquely determines the transition probability of X_t, is defined in accordance with (2.4.10) as the uniform limit

$$(9.4.3) \qquad A g (s, x) = \lim_{t \downarrow 0} \frac{E g (t + s, X_{t+s} (s, x)) - g (s, x)}{t},$$

where $g (s, x)$ is a bounded measurable function defined on $[t_0, T] \times R^d$.

For the moment, we neglect the requirement of uniformity of the limit and seek to ascertain when at least the pointwise limit in (9.4.3) exists. If we call the result L, we have

$$(9.4.4) \qquad L g = \frac{\partial g}{\partial s} + \mathfrak{D} g,$$

which holds for all functions g defined on $[t_0, T] \times R^d$ that have continuous first partial derivatives with respect to t and continuous second partial derivatives with respect to the components of x and that, together with their derivatives, do not, as a function of x, increase faster than some fixed power of x. This is easily seen from (9.4.3) by replacing $g (t + s, X_{t+s} (s, x))$ in that expression with the stochastic integral of Itô's theorem, namely (for brevity, we write X_{t+s} for $X_{t+s} (s, x)$),

$$g (t + s, X_{t+s}) = g (s, x) + \int_s^{t+s} g_s (u, X_u) \, du + \sum_{i=1}^d \int_s^{t+s} f_i (u, X_u) \, g_{x_i} (u, X_u) \, du$$

$$+ \frac{1}{2} \sum_{i=1}^d \sum_{j=1}^d \int_s^{t+s} b_{ij} (u, X_u) \, g_{x_i} g_{x_j} (u, X_u) \, du$$

$$+ \sum_{i=1}^d \sum_{k=1}^m \int_s^{t+s} g_{x_i} (u, X_u) \, G_{ij} (u, X_u) \, dW_u^j,$$

and taking the limit. If g depends only on x, we have in the homogeneous case, in place of (9.4.4),

$$L = \mathfrak{D}.$$

The limit in (9.4.3) exists and is uniform if g has the above-listed properties and vanishes identically outside a bounded subset of $[t_0, T] \times R^d$. Such functions therefore, belong to the domain of definition D_A of A. For them,

$$A g = \frac{\partial g}{\partial s} + \mathfrak{D} g,$$

and, in the case of homogeneous processes,

$$A g = \mathfrak{D} g.$$

Thus, we have found the form of the infinitesimal operator of the solution of the stochastic differential equation (9.4.1). In this case, the solution is uniquely determined by f and G.

In remark (2.6.5), we showed how the transition probabilities of a diffusion process can be found in theory from knowledge of the functions

$$u(s, x) = E g(X_t(s, x)),$$

where t is fixed and g ranges over a set of functions that is dense in the space $C(R^d)$ of continuous bounded functions defined on R^d. For given g, we can calculate $u(s, x)$ from Kolmogorov's backward equation. This equation is valid here under the following assumptions:

(9.4.4)   Theorem. Suppose that the assumptions of Theorem (9.3.1) are satisfied for equation (9.4.1). Suppose also that the coefficients f and G have continuous bounded first and second partial derivatives with respect to the components of x. Then, if $g(x)$ is a continuous bounded function with continuous bounded first and second partial derivatives, the function

$$u(s, x) = E g(X_t(s, x)), \quad t_0 \leqq s \leqq t \leqq T, \quad x \in R^d,$$

and its first and second partial derivatives with respect to x and its first derivative with respect to s are continuous and bounded. Also, the *backward equation*

$$(9.4.5) \qquad \frac{\partial u(s, x)}{\partial s} + \mathfrak{D} u(s, x) = 0,$$

where $\mathfrak{D}$ is the differential operator (9.4.2), with the end condition

$$\lim_{s \uparrow t} u(s, x) = g(x)$$

is valid.

The proof of these assertions follows, on the basis of remark (7.3.7), from Theorem (2.6.3).

Instead of solving the backward equation (9.4.5) for a set of end values g that is dense in $C(R^d)$, we can confine our attention to the family

$$g(x) = e^{i\lambda' x}, \quad \lambda \in R^d .$$

We obtain

$$u(s, x) = E \exp(i \lambda' X_t(s, x)),$$

that is, the characteristic function of $X_t(s, x)$, which determines uniquely the probability distribution of $X_t(s, x)$, namely, $P(s, x, t, \cdot)$.

If $P(s, x, t, B)$ has a density $p(s, x, t, y)$, we can get equations for $p(s, x, t, y)$ itself from Theorem (2.6.6) and (2.6.9). The density is a fundamental solution of the backward equation; that is, for fixed t and y and for $s < t$,

$$\frac{\partial}{\partial s} p\,(s, x, t, y) + \sum_{i=1}^{d} f_i\,(s, x) \frac{\partial}{\partial x_i} p\,(s, x, t, y)$$

(9.4.6)

$$+ \frac{1}{2} \sum_{i=1}^{d} \sum_{j=1}^{d} b_{ij}\,(s, x) \frac{\partial^2}{\partial x_i\,\partial x_j} p\,(s, x, t, y) = 0,$$

$$\lim_{s \uparrow t} p\,(s, x, t, y) = \delta\,(y - x),$$

just as it is a fundamental solution of the *forward* or *Fokker-Planck equation*; that is, for s and x fixed and $t > s$,

$$\frac{\partial}{\partial t} p\,(s, x, t, y) + \sum_{i=1}^{d} \frac{\partial}{\partial y_i} (f_i\,(t, y)\, p\,(s, x, t, y))$$

(9.4.7)

$$- \frac{1}{2} \sum_{i=1}^{d} \sum_{j=1}^{d} \frac{\partial^2}{\partial y_i\,\partial y_j} (b_{ij}\,(t, y)\, p\,(s, x, t, y)) = 0,$$

$$\lim_{t \downarrow s} p\,(s, x, t, y) = \delta\,(y - x).$$

However, these laws of development for p are valid only under certain assumptions regarding the coefficients f and $B = G\,G'$ (see section 2.6).

We refer to a theorem of Gikhman and Skorokhod ([36], pp. 96-99) that, for the scalar case, gives sufficient conditions for existence of a density with certain analytical properties.

(9.4.8) **Example.** In accordance with section 8.2, the scalar autonomous linear stochastic differential equation $(d = m = 1)$

$$dX_t = (A\,X_t + a)\,dt + b\,dW_t, \quad X_0 = c, \quad t \geqq 0,$$

has the solution

$$X_t = c\,e^{At} + \frac{a}{A}\,(e^{At} - 1) + b \int_0^t e^{A(t-s)}\,dW_s.$$

Special cases are the Ornstein-Uhlenbeck process $(a = 0)$, the deterministic linear equation with random initial value $(b = 0)$, and the Wiener process $(A = a = 0,\ b = 1,\ c = 0)$. For $b = 0$, X_t has a density if c has a density, but the transition probabilities degenerate to

$$P\,(s, x, t, \cdot) = \delta_z, \quad z = x\,e^{A(t-s)} + \frac{a}{A}\,(e^{A(t-s)} - 1)\ .$$

For $b \neq 0$, there exists a density $p\,(s, x, t, y)$ so smooth that it can be obtained from the backward equation

$$\frac{\partial}{\partial s} p(s, x, t, y) + (A x + a) \frac{\partial}{\partial x} p(s, x, t, y) + \frac{1}{2} b^2 \frac{\partial^2}{\partial x^2} p(s, x, t, y) = 0$$

or from the forward equation

$$\frac{\partial}{\partial t} p(s, x, t, y) + \frac{\partial}{\partial y} ((A y + a) p(s, x, t, y)) - \frac{1}{2} \frac{\partial^2}{\partial y^2} (b^2 p(s, x, t, y)) = 0$$

as the fundamental solution. As boundary condition, we assume that p and its partial derivatives with respect to x_i and y_i vanish respectively as $|x| \to \infty$ and $|y| \to \infty$. We know from example (9.2.12) that

$$p(s, x, t, y) = (2 \pi K_t(s, x))^{-1/2} \exp\left(-(y - m_t(s, x))^2 / 2 K_t(s, x)\right)$$

where

$$m_t(s, x) = x e^{A(t-s)} + \frac{a}{A} (e^{A(t-s)} - 1)$$

and

$$K_t(s, x) = \frac{b^2}{2 A} (e^{2 A(t-s)} - 1),$$

that is, $p(s, x, t, y)$ is the density of $\mathfrak{N}(m_t(s, x), K_t(s, x))$. In the special case $A = 0$, we have

$$m_t(s, x) = x + a(t - s),$$

$$K_t(s, x) = b^2(t - s).$$

We observe that $p(s, x, t, y)$ depends only on $t - s$. Thus, X_t is a homogeneous diffusion process, as has to be the case for the solution of an autonomous equation.

The backward and forward equations (9.4.6) and (9.4.7) have up to now been explicitly solved only in a few simple cases (see, for example, Uhlenbeck and Ornstein [49], Wang and Uhlenbeck [50], or Bharucha-Reid [19]). These have usually been found by taking the Fourier or Laplace transformation of p.

(9.4.9) **Example.** The forward equation for the general linear equation

$$dX_t = (a(t) + A(t) X_t) dt + \sum_{i=1}^{m} (B_i(t) X_t + b_i(t)) dW_t^i$$

with drift vector

$$f(t, x) = a(t) + A(t) x, \quad a = (a^1, \ldots, a^d)', \quad A = (A_{ij}),$$

and diffusion matrix

$$B(t, x) = \sum_{i=1}^{m} (B_i \, x \, x' \, B_i' + B_i \, x \, b_i' + b_i \, x' \, B_i' + b_i \, b_i')$$

(see example (9.3.4)) becomes, for the density $p(s, x, t, y)$ and $t > s$

$$\frac{\partial}{\partial t} p(s, x, t, y) + \sum_{i=1}^{d} a^i(t) \frac{\partial}{\partial y_i} p(s, x, t, y) + \sum_{i=1}^{d} \sum_{j=1}^{d} A_{ij}(t) \frac{\partial}{\partial y_i} (y_j \, p)$$

$$- \frac{1}{2} \sum_{i=1}^{d} \sum_{j=1}^{d} \frac{\partial^2}{\partial y_i \, \partial y_j} (b_{ij}(t, y) \, p(s, x, t, y)) = 0,$$

with the initial condition

$$\lim_{t \downarrow s} p(s, x, t, y) = \delta(y - x).$$

For the case $B_1(t) = \ldots = B_m(t) \equiv 0$, in the notation

$$G(t) = (b_1(t), \ldots, b_m(t)),$$

$$\frac{\partial}{\partial t} p = p_t, \quad \left(\frac{\partial}{\partial y_1} p, \ldots, \frac{\partial}{\partial y_d} p \right)' = p_y, \quad \left(\frac{\partial^2}{\partial y_i \, \partial y_j} p \right) = p_{yy}$$

we obtain, more specially,

$$(9.4.10) \qquad p_t + a(t)' \, p_y + p \, \text{tr}(A(t)) + p_y' \, A(t) \, y - \frac{1}{2} \, \text{tr}(G(t) \, G(t)' \, p_{yy}) = 0.$$

We know from example (9.2.12) that the solution of this equation is the density of a normal distribution whose parameters $m_t(s, x)$ and $K_t(s, x)$ are given in (9.2.12). The dynamic development of these parameters is characterized by the differential equations of Theorem (8.2.6).

(9.4.11) **Remark.** The distribution of an R^p-valued functional $g(X_t)$ dependent only on the state X_t at the instant t can always be obtained via the characteristic function

$$u(s, x) = E \, e^{i \lambda' g(X_t(s, x))}, \quad \lambda \in R^p$$

from the backward equation (9.4.5) with the end condition

$$u(t, x) = e^{i \lambda' g(x)}$$

Many interesting quantities (for example, the time of first entry or the length of stay in specific regions) depend, however, on the overall course of a trajectory in a time interval. But, in certain cases, one can give a differential equation for their characteristic function. For example, if $g(x)$ is R^p-valued and $h(t, x)$ is R^q-valued, put

$$V_{\lambda,\mu}(s,x) = E_{s,x} \exp\left(i\,\lambda'\,g\,(X_t) + i\,\mu'\int\limits_{s}^{t} h\,(u,X_u)\,\mathrm{d}u\right)$$

for $s < t$, $\lambda \in R^p$, and $\mu \in R^q$. Under certain conditions on f, G, g, and h (see Gikhman and Skorokhod [5], p. 414 or [36], p. 302), we have the equation

$$\frac{\partial}{\partial s}\,V_{\lambda,\mu}(s,x) + \mathfrak{D}\,V_{\lambda,\mu}(s,x) + i\,\mu'\,h\,(s,x)\,V_{\lambda,\mu}(s,x) = 0$$

with the end condition

$$\lim_{s \uparrow t} V_{\lambda,\mu}(s,x) = e^{i\,\lambda'\,g\,(x)}.$$

Here, $\mathfrak{D}$ is the operator (9.4.2).

(9.4.12) **Remark.** Diffusion processes whose state X_t can assume only values in a subset of R^d the boundary of which can be absorbing, reflecting, or elastic have been discussed, for example, by Gikhman and Skorokhod [5], Prohorov and Rozanov [15], Dynkin [21], Itô and McKean [26], and Mandl [28].

Chapter 10
Questions of Modelling and Approximation

10.1 The Shift From a Real to a Markov Process

Many continuous dynamic systems under the influence of a random disturbance can be represented by an ordinary differential equation (in general nonlinear) of the form

$$(10.1.1) \quad \dot{X}_t = f(t, X_t, Y_t), \quad t \geq t_0, \quad X_{t_0} = c,$$

Here, X_t is the d-dimensional state vector and Y_t is an m-dimensional disturbance process whose probability-theoretic characteristics (finite-dimensional distributions) we assume to be given. We also assume that the distribution of the initial value c is given. Now if Y_t is a stochastic process with sufficiently smooth (e.g. continuous) sample functions, (10.1.1) can be considered as an ordinary differential equation for the sample functions $X.(\omega)$ of the state of the system.

However, the solution process X_t is a Markov process only if, for known $X_t = x$, the value of X_{t+h} given approximately by

$$X_{t+h} = x + f(t, x, Y_t) h$$

is independent of what took place prior to the instant t, that is, only if Y_t is a process with statistically independent values at every point. This is the case, for example, when Y_t does not depend at all on chance but is equal to a fixed function, especially if $Y_t \equiv 0$ does not appear. Then, (10.1.1) degenerates into a deterministic equation.

However, if Y_t is a nondegenerate stochastic process with independent values at every point, for example, a stationary Gaussian process such that $E\,Y_t \equiv 0$ and

$$E\,Y_t\,Y_s' = \begin{cases} 0 & \text{for } t \neq s, \\ I & \text{for } t = s, \end{cases}$$

the sample functions of Y_t are extremely irregular; for example, they are discontinuous everywhere and unbounded on every interval, and their graphs are "point clouds" dense in $[t_0, \infty) \times R^m$. If we approximate such a process with a sequence

$Y_t^{(n)}$ of continuous stationary Gaussian processes such that $E\,Y_t^{(n)} \equiv 0$ and $E\,Y_t^{(n)}\,Y_s^{(n)\prime} = e^{-n|t-s|}\,I$, we obtain

$$E\left(\int_{t_0}^t Y_s^{(n)}\,\mathrm{d}s\right) = 0, \quad E\left|\int_{t_0}^t Y_s^{(n)}\,\mathrm{d}s\right|^2 \to 0 \quad (n \to \infty).$$

Thus, the disturbing action of the process $Y_t^{(n)}$ has, on the (timewise) average, less and less effect on the left-hand member of (10.1.1) as $n \to \infty$ since the variance $E\,|Y_t^{(n)}|^2 = d$ (that is, the average energy of the disturbance) remains finite.

Thus, we are led inexorably to rapidly fluctuating processes with infinite energy and hence to generalized stochastic processes with independent values at every point (in the sense of section 3.2). So-called "delta-correlated" Gaussian processes are examples of this.

We now need to confine ourselves, for a meaningful theory, to functions $f(t, x, y)$ that are *linear* in y, so that (10.1.1) has the form

$$\dot{X}_t = f(t, X_t) + G(t, X_t)\,Y_t\;.$$

For it is still possible to define the product of an ordinary and a generalized function whereas the square of a generalized function (for example, $\delta(t)\,\delta(t)$) is no longer in general defined.

In accordance with the remark at the end of section 3.2, we can without loss of generality confine ourselves to white noise ξ_t as the prototype of a "delta-correlated" Gaussian noise process, thus considering only equations of the form

(10.1.2) $\dot{X}_t = f(t, X_t) + G(t, X_t)\,\xi_t$

for which we have, with the aid of the stochastic integral, developed a precise mathematical theory. The solution of (10.1.2) is a Markov process and hence belongs to a class of processes for the analysis of which efficient mathematical methods exist though it has, on the other hand, the disadvantage that its sample functions are not smooth functions (see section 7.2). This last fact is typical of Markov processes since the Markov property, formulated in a *negative* way, states that, for a known present, it is forbidden to transmit information from the past into the future. The "jagged" behavior of X_t arises from this.

Now, physically realizable processes are always smooth processes and hence are at best only approximately Markov processes. In equation (10.1.2), for example, we would have in actuality not the white noise ξ_t but only an approximate "delta-correlated" process ζ_t, hence not a white but a "colored" noise in a general sense.

Let us look at the scalar equation

(10.1.3) $\dot{X}_t = f(t, X_t) + G(t, X_t)\,\zeta_t,$

where ζ_t is the stationary Ornstein-Uhlenbeck process, hence the solution of the stochastic differential equation

$$\dot{\zeta}_t = -\alpha\,\zeta_t + \sigma\,\xi_t$$

(see section 8.3) with $\alpha > 0$ and $\Re\,(0, \sigma^2/2\,\alpha)$-distributed initial value. The process ζ_t has the covariance function

$$E\,\zeta_t\,\zeta_s = e^{-\alpha\,|t-s|}\,\sigma^2/2\,\alpha.$$

Therefore, in accordance with section 3.2, it becomes an approximation of white noise as $\alpha \to \infty$, $\sigma^2 \to \infty$, but $\sigma^2/2\,\alpha^2 = D \to 1/2$. By virtue of the continuity of ζ_t, equation (10.1.3) can be regarded as an ordinary differential equation. Hence, for sufficiently smooth f and G, it provides a process X_t, which is now differentiable though not a Markov process. However, we can regain the Markov property by shifting to a two-dimensional state space and considering the vector process (X_t, ζ_t). The replacement of ξ_t with a stationary Gaussian process ζ_t possessing analytical sample functions is discussed in Stratonovich [76], pp. 125-126. The *basic rule is* as follows: For every newly attained derivative of X_t, we need (in the scalar case) to add a component to the accompanying Markov process.

The question as to what stochastic differential equation must now be chosen in order to describe adequately a given physically realizable process is the question of **modelling**. A controversy has arisen in this connection (see Gray and Gaughey [39] and McShane [46]) since different authors have obtained different solutions for apparently identical problems. These discrepancies arise, as we shall see, not from errors in the mathematical calculation but from a general discontinuity of the relationship between differential equations for stochastic processes and their solutions. Let us clarify this with an example.

(10.1.4)   **Example.** By virtue of Corollary (8.4.3a), the solution of the scalar stochastic differential equation

$$\dot{X}_t = A\,(t)\,X_t + B\,(t)\,X_t\,\xi_t, \quad X_{t_0} = c \in L^2,$$

where ξ_t is a scalar white noise is, after we have written it in the form

$$\mathrm{d}X_t = A\,(t)\,X_t\,\mathrm{d}t + B\,(t)\,X_t\,\mathrm{d}W_t, \quad X_{t_0} = c\,,$$

the process

$$X_t = c\,\exp\left(\int\limits_{t_0}^{t}(A\,(s) - B\,(s)^2/2)\,\mathrm{d}s + \int\limits_{t_0}^{t} B\,(s)\,\mathrm{d}W_s\right).$$

Let us now replace in the original equation the white noise ξ_t with a sequence of physically realizable continuous Gaussian stationary processes $\{\xi_t^{(n)}\}$ such that $E\,\xi_t^{(n)} \equiv 0$ and $E\,\xi_t^{(n)}\,\xi_s^{(n)} = C_n\,(t-s)$, where

$$\lim_{n \to \infty} C_n\,(t) = \delta\,(t).$$

Then,

$$\dot{Y}_t = A(t) Y_t + B(t) Y_t \xi_t^{(n)}, \quad X_{t_0} = c,$$

is an ordinary differential equation and hence has the solution

$$Y_t^{(n)} = c \exp\left(\int_{t_0}^{t} A(s)\, ds + \int_{t_0}^{t} B(s)\, \xi_s^{(n)}\, ds\right).$$

The process

$$Z_t^{(n)} = \int_{t_0}^{t} B(s)\, \xi_s^{(n)}\, ds$$

is a Gaussian process with mean 0 and with covariance

$$E Z_t^{(n)} Z_s^{(n)} = \int_{t_0}^{t} \int_{t_0}^{s} B(u)\, B(v)\, C_n(u-v)\, du\, dv$$

$$\longrightarrow \int_{t_0}^{\min(t,s)} B(u)^2\, du,$$

that is, $Y_t^{(n)}$ converges in mean square to a process whose distributions coincide with the distributions of the process

$$Y_t = c \exp\left(\int_{t_0}^{t} A(s)\, ds + \int_{t_0}^{t} B(s)\, dW_s\right).$$

Therefore, the processes X_t and Y_t are quite different for $B(t) \not\equiv 0$. From Corollary (8.4.3a), Y_t is the solution of the stochastic differential equation

$$dY_t = (A(t) + B(t)^2/2)\, Y_t\, dt + B(t)\, Y_t\, dW_t, \quad Y_{t_0} = c.$$

To get X_t, we made the limiting shift to the white noise ξ_t *in the original equation* and solved the equation as a stochastic differential equation. In contrast, we obtained Y_t by solving the ordinary differential equation disturbed by $\xi_t^{(n)}$ and then made this shift to ξ_t *in the solution*. Obviously, this leads to different results. Both processes (though not the processes $Y_t^{(n)}$) are Markov processes since they satisfy (different!) stochastic differential equations. Which of these is the "correct" process (in the sense of giving the better description of the basic system) can in general be decided only *pragmatically*.

If we denote by $L(g)$ the solution of an equation g and by $g(\xi_t)$ or $g(\xi_t^{(n)})$ the stochastic differential equation

$$\dot{X}_t = f + G\, \xi_t, \quad \xi_t \text{ is a white noise},$$

or its approximating ordinary differential equation

$$\dot{X}_t = f + G\,\xi_t^{(n)}, \quad \xi_t^{(n)} \text{ is a continuous stationary Gaussian process,}$$

$$E\,\xi_t^{(n)} = 0, \quad E\,\xi_t^{(n)}\,\xi_s^{(n)} = C_n\,(t-s),$$

respectively, then example (10.1.4) shows that, in general,

$$L\,(g\,(\xi_t)) \neq \lim_{C_n(t)\to\delta(t)} L\,(g\,(\xi_t^{(n)}))\ .$$

The question now arises as to whether we can modify the definition of the stochastic integral in such a way that equality will hold in the last relationship. This is in fact possible—by means of the definition (already mentioned in section 4.2) of Stratonovich's time-symmetric stochastic integral [48] .

10.2 Stratonovich's Stochastic Integral

In section 4.2, we pointed out that, in the attempt to evaluate the integral

$$\int_{t_0}^{t} W_s\,dW_s$$

as the limiting value of the approximating sums

$$S_n = \sum_{i=1}^{n} W_{\tau_i}\,(W_{t_i} - W_{t_{i-1}}), \quad t_0 \leq t_1 \leq \ldots \leq t_n = t,\ t_{i-1} \leq \tau_i \leq t_i\,,$$

the result depends very much on the choice of intermediate points τ_i. Itô's choice $\tau_i = t_{i-1}$, on which we have based our exposition exclusively up to now led to the value

$$\int_{t_0}^{t} W_s\,dW_s = \underset{\delta_n \to 0}{\text{qm-}\lim}\ S_n = (W_t^2 - W_{t_0}^2)/2 - (t - t_0)/2$$

where

$$\delta_n = \max\,(t_i - t_{i-1})$$

and, in general, to a concept of an integral that is not symmetric with respect to the variable t since the increments dW_s "point into the future".

However, it is just this lack of symmetry that leads to the simple formulas for the first two moments of the integral (see Theorem (4.4.14e)) and to the martingale property (see Theorem (5.1.1b)). Furthermore, a stochastic differential equation

$$dX_t = f(t, X_t)\, dt + G(t, X_t)\, dW_t$$

explained on this basis yields, under the assumptions of Theorem (9.3.1), a diffusion process as solution. The intuitive significance of the coefficients f and G is explained by regarding f as the drift and $G\,G'$ as the diffusion matrix of that process. A disadvantage is that the calculus valid for stochastic differential equations, which operates in accordance with Itô's theorem, deviates from the familiar one.

This disadvantage (together with all the advantages of Itô's integral that we have mentioned) is removed by Stratonovich's definition [48], which yields for the special case considered at the beginning

$$(S) \int_{t_0}^{t} W_s\, dW_s = \underset{\delta_n \to 0}{\text{qm-lim}} \sum_{i=1}^{n} \frac{W_{t_{i-1}} + W_{t_i}}{2} (W_{t_i} - W_{t_{i-1}})$$

$$= (W_t^2 - W_{t_0}^2)/2$$

hence a value that we can also obtain by formal integration by parts.

Somewhat more generally, let us define

$$(10.2.1) \quad (S) \int_{t_0}^{t} H(s, W_s)\, dW_s = \underset{\delta_n \to 0}{\text{qm-lim}} \sum_{i=1}^{n} H\left(t_{i-1}, \frac{W_{t_{i-1}} + W_{t_i}}{2}\right)(W_{t_i} - W_{t_{i-1}}).$$

Here, W_t is an m-dimensional Wiener process and $H(t, x)$ is a $(d \times m$ matrix)-valued function that is continuous with respect to t, that has first-order partial derivatives H_{x_j} with respect to the m components x_j of x, and that satisfies the condition

$$\int_{t_0}^{t} E\, |H(s, W_s)|^2\, ds < \infty \ .$$

It follows from Theorem (10.2.5) that the limit in (10.2.1) exists. The result is connected with Itô's integral, which is defined in the present case by

$$\int_{t_0}^{t} H(s, W_s)\, dW_s = \underset{\delta_n \to 0}{\text{qm-lim}} \sum_{i=1}^{n} H(t_{i-1}, W_{t_{i-1}})(W_{t_i} - W_{t_{i-1}})$$

as shown in the following equation:

$$(10.2.2) \quad (S) \int_{t_0}^{t} H(s, W_s)\, dW_s = \int_{t_0}^{t} H(s, W_s)\, dW_s + \frac{1}{2} \int_{t_0}^{t} \sum_{k=1}^{m} (H_{x_k}(s, W_s))_{\cdot k}\, ds.$$

Here, the d-vector $(H_{x_k})_{\cdot k}$ is the kth column of the $d \times m$ matrix H_{x_k}.

To build a theory of stochastic differential equations on the basis of Stratono-vich's idea, we use the following general

(10.2.3) Definition. Let Y_t denote an m-dimensional diffusion process on the interval $[t_0, T]$ for which relations b) and c) in definition (2.1.5) hold for $\varepsilon = \infty$ and whose drift vector $a\,(t, x)$ and diffusion matrix $B\,(t, x)$ together with the derivatives $\partial B\,(t, x)/\partial x_j$, for $j = 1, \dots, m$, are continuous in both arguments. Suppose also that $H\,(t, x)$ is a $(d \times m$ matrix$)$-valued function that is continuous in t, that has continuous partial derivatives $\partial H\,(t, x)/\partial x_j$, and that satisfies, for $t \in [t_0, T]$, the conditions

$$\int\limits_{t_0}^{t} E\,|H\,(s, Y_s)\,a\,(s, Y_s)|\,ds < \infty$$

and

$$\int\limits_{t_0}^{t} E\,|H\,(s, Y_s)\,B\,(s, Y_s)\,H\,(s, Y_s)'|\,ds < \infty \quad .$$

Then, the limiting value

$$(10.2.4) \quad (S) \int\limits_{t_0}^{t} H\,(s, Y_s)\,dY_s = \operatorname*{qm-lim}_{\delta_n \to 0} \sum_{i=1}^{n} H\left(t_{i-1}, \frac{Y_{t_{i-1}} + Y_{t_i}}{2}\right)(Y_{t_i} - Y_{t_{i-1}}),$$

where $t_0 \leqq t_1 \leqq \dots \leqq t_n = t$ is a partition of the interval $[t_0, t]$ and $\delta_n = \max (t_i - t_{i-1})$, is called the **stochastic integral in the sense of Stratonovich.**

This integral is connected with Itô's integral as follows:

(10.2.5) Theorem. The limit in (10.2.4) exists under the conditions mentioned in definition (10.2.3). It is connected with Itô's stochastic integral, defined here by

$$(10.2.6) \quad \int\limits_{t_0}^{t} H\,(s, Y_s)\,dY_s = \operatorname*{qm-lim}_{\delta_n \to 0} \sum_{i=1}^{n} H\,(t_{i-1}, Y_{t_{i-1}})(Y_{t_i} - Y_{t_{i-1}})$$

by the formula

$$(S)\int\limits_{t_0}^{t} H(s,Y_s)\,dY_s = \int\limits_{t_0}^{t} H(s,Y_s)\,dY_s + \frac{1}{2}\sum_{j=1}^{m}\sum_{k=1}^{m}\int\limits_{t_0}^{t}(H_{x_k}(s,Y_s))_{\cdot j}\,b_{jk}(s,Y_s)\,ds \quad .$$

Here, the d-vector $(H_{x_k})_{\cdot j}$ is the jth column of the $d \times m$ matrix $H_{x_k} = (\partial H_{ij}/\partial x_k)$.

To prove this, we consider the difference between the sums in (10.2.4) and (10.2.6):

$$\sum_{i=1}^{n} \left(H\left(t_{i-1}, \frac{Y_{t_{i-1}}+Y_{t_i}}{2}\right) - H\left(t_{i-1}, Y_{t_{i-1}}\right)\right) (Y_{t_i} - Y_{t_{i-1}}) \ .$$

We then apply the mean-value theorem to the terms $H\left(t_{i-1}, (Y_{t_{i-1}} + Y_{t_i})/2\right)$. For details, see Stratonovich [48].

(10.2.7) **Remark.** For $d = m = 1$, the conversion formula is

$$\text{(S)} \int_{t_0}^{t} H(s, Y_s)\, dY_s = \int_{t_0}^{t} H(s, Y_s)\, dY_s + \frac{1}{2} \int_{t_0}^{t} \frac{\partial H(s, Y_s)}{\partial x} B(s, Y_s)\, ds.$$

We can see from this, in particular, that the stochastic integrals of Itô and Stratonovich coincide when $H(t, x) \equiv H(t)$ is independent of x.

We can now define the stochastic differential equations of the customary form in terms of the integral

$$\text{(S)} \int_{t_0}^{t} (0, G(s, X_s))\, d\binom{X_s}{W_s} = \text{(S)} \int_{t_0}^{t} G(s, X_s)\, dW_s$$

(where X_t is d-dimensional and W_t an m-dimensional Wiener process and $G(t, x)$ is $(d \times m$ matrix)-valued) and the equation

$$X_t = X_{t_0} + \int_{t_0}^{t} f(s, X_s)\, ds + \text{(S)} \int_{t_0}^{t} G(s, X_s)\, dW_s, \quad t_0 \leqq t \leqq T.$$

We again write symbolically

$$\text{(S)} \ dX_t = f(t, X_t)\, dt + G(t, X_t)\, dW_t,$$

where the parenthesized S on the left means that the stochastic integral on the basis of which the differential notation is defined is to be understood in the sense of Stratonovich.

For our case, where $H = (0, G)$ and $Y_t = (X_t, W_t)$, the conversion formula of Theorem (10.2.5) yields

$$\text{(S)} \int_{t_0}^{t} G(s, X_s)\, dW_s = \int_{t_0}^{t} G(s, X_s)\, dW_s + \frac{1}{2} \sum_{j=1}^{m} \sum_{k=1}^{d} \int_{t_0}^{t} (G_{x_k}(s, X_s))_{\cdot j}$$

$$G_{kj}(s, X_s)\, ds$$

(see Stratonovich [48]), so that the Itô equation

$$dX_t = \left(f(t, X_t) + \frac{1}{2} \sum_{j=1}^{m} \sum_{k=1}^{d} (G_{x_k}(t, X_t))_{\cdot j} \, G_{kj}(t, X_t) \right) dt + G(t, X_t)\, dW_t$$

corresponds to the above-defined Stratonovich differential equation in the sense of coincidence of solutions. Conversely, the Stratonovich equation

$$\text{(S)} \quad dX_t = \left(f - \frac{1}{2} \sum_{j=1}^{m} \sum_{k=1}^{d} (G_{x_k})_{.j} \, G_{kj} \right) dt + G \, dW_t$$

corresponds to the Itô equation

$$(10.2.8) \quad dX_t = f \, dt + G \, dW_t \, .$$

Thus, whether we think of a given formal equation (10.2.8) in the sense of Itô or Stratonovich, we arrive at the same solution as long as $G(t, x) \equiv G(t)$ is independent of x, as in the case of an equation that is linear in the narrow sense (see section 8.2). In general, we obtain two distinct Markov processes as solutions, which differ in the systematic (drift) behavior but not in the fluctuational behavior. This last means, in particular, that the sample functions of the solution of a Stratonovich equation are also in general functions that are not differentiable or of bounded variation.

(10.2.9) Example. The formal scalar linear equation

$$(?) \quad dX_t = A(t) X_t \, dt + B(t) X_t \, dW_t, \quad X_{t_0} = c \in L^2,$$

taken as an Itô equation has the solution

$$X_t = c \exp \left(\int_{t_0}^{t} (A(s) - B(s)^2/2) \, ds + \int_{t_0}^{t} B(s) \, dW_s \right),$$

but taken as a Stratonovich equation has the solution

$$Y_t = c \exp \left(\int_{t_0}^{t} A(s) \, ds + \int_{t_0}^{t} B(s) \, dW_s \right).$$

One should compare the example (10.1.4). We can also obtain Y_t by formal solution of the original equation. In general, the two processes have quite different global properties. For example, if we set $A(t) \equiv A$, $B(t) \equiv B$, and $t_0 = 0$, then

$$X_t = c \, e^{(A - B^2/2)t + BW_t} \longrightarrow 0 \quad \text{(almost certainly)}$$

as $t \longrightarrow \infty$ if and only if $A < B^2/2$ whereas

$$Y_t = c \, e^{At + BW_t} \longrightarrow 0 \quad \text{(almost certainly)}$$

as $t \longrightarrow \infty$ if and only if $A < 0$.

It can be shown in general (see Stratonovich [48]) that the Stratonovich integral and the Stratonovich differential defined in terms of it satisfy all the formal rules of an ordinary integral or differential (integration by parts, change of variable, chain rule) and hence in this respect can "more easily" be manipulated than the Itô integral or differential. Unfortunately, the price we have to pay for this is the loss of all the advantages of Itô's integral that were mentioned earlier. However, the conversion formulas that we have given enable us at all times to shift from one type of integral to the other.

The system-theoretic significance of Stratonovich equations consists in the fact that, in many cases, they present themselves automatically when one approximates a white noise or a Wiener process with smoother processes, solves the approximating equation, and in the solution shifts back to the white noise. Comparison of (10.1.4) and (10.2.9) shows this immediately. In the following section, we shall discuss this matter in greater detail.

10.3 Approximation of Stochastic Differential Equations

Suppose that

$$(10.3.1) \quad \dot{X}_t = f(t, X_t) + G(t, X_t) \, \zeta_t, \quad t_0 \leq t \leq T, \quad X_{t_0} = c,$$

is a formal differential equation for the d-dimensional state of a dynamic system. In this equation, let us assume that $f(t, x) \in R^d$, that $G(t, x)$ is ($d \times m$ matrix)-valued, and that the m-dimensional process ζ_t is a "rapidly fluctuating" stationary Gaussian process that is independent of c and has expectation 0. How should equation (10.3.1) be interpreted?

If we know or can assume that ζ_t is exactly or nearly equal to a white noise process ξ_t, (10.3.1) should be interpreted as the Itô differential equation

$$(10.3.2) \quad dX_t = f(t, X_t) \, dt + G(t, X_t) \, dW_t, \quad X_{t_0} = c.$$

The same is true if (10.3.1) serves as an approximation or a limit of a discrete (timewise nonsymmetric) problem

$$\frac{X_{t_{k+1}} - X_{t_k}}{t_{k+1} - t_k} = f(t_k, X_{t_k}) + G(t_k, X_{t_k}) \, \zeta_{t_k}$$

where the ζ_{t_k} are Gaussian, independent, and identically distributed.

On the other hand, if ζ_t is a continuous process and only an approximation of the white noise (for example, with the delta-like covariance function $C(t)$), we can treat (10.3.1) as an ordinary differential equation and solve it by the classical procedures. The solution is not a Markov process, but under certain conditions (see Gray [38] or Clark [33]) it converges in mean square, as

$$C(t) \rightarrow \delta(t)$$

or

$$\text{qm-}\lim \int_{t_0}^{t} \zeta_s \, ds = W_t$$

to a Markov process, which is now the solution of the corresponding Stratonovich equation

(10.3.3)   (S) $dX_t = f(t, X_t) \, dt + G(t, X_t) \, dW_t, \quad X_{t_0} = c$.

One should again compare examples (10.1.4) and (10.2.9).

We emphasize once again that the solutions of equations (10.3.2) and (10.3.3) coincide when $G(t, x) \equiv G(t)$ is independent of x.

Suppose that $t_0 < t_1 < \dots < t_n = T < \infty$ is a decomposition of $[t_0, T]$ and that $\delta_n = \max(t_{k+1} - t_k)$. Khas'minskiy ([65], pp. 218-220) points out that, as $\delta_n \to 0$, the sequence of "Cauchy polygonal lines" $X_t^{(n)}$ defined by

$$X_{t_0} = c \in L^2,$$
$$X_{t_{k+1}}^{(n)} = X_{t_k}^{(n)} + f(t_k, X_{t_k}^{(n)})(t_{k+1} - t_k) + G(t_k, X_{t_k}^{(n)})(W_{t_{k+1}} - W_{t_k})$$

and by linear interpolation between the partition points converges in mean square to the solution of the Itô equation (10.3.2). If we set

$$X_{t_{k+1}}^{(n)} = X_{t_k}^{(n)} + f(t_k, X_{t_k}^{(n)})(t_{k+1} - t_k) + G\left(t_k, \frac{X_{t_k}^{(n)} + X_{t_{k+1}}^{(n)}}{2}\right)(W_{t_{k+1}} - W_{t_k})$$

this leads to the solution of Stratonovich's equation (10.3.3).

Since a stochastic differential equation of the form (10.3.2) is merely a short way of writing the integral equation

$$X_t = c + \int_{t_0}^{t} f(s, X_s) \, ds + \int_{t_0}^{t} G(s, X_s) \, dW_s, \quad t_0 \leq t \leq T,$$

it seems natural to try to construct an approximation for X_t by replacing W_t in the second integral with a smooth integrator, so that the integral can be regarded as an ordinary Riemann-Stieltjes integral and can be evaluated. We again see that a smoothing *prior* to evaluation of the stochastic integral leads, when we take the limit in the result, to the solution of the formally identical Stratonovich equation.

For example, we may approximate the trajectories $W_{\cdot}(\omega)$ with a sequence $W_{\cdot}^{(n)}(\omega)$ of continuous functions of bounded variation and piecewise-continuous derivatives, so that, for all $a \in [t_0, T]$ and almost all $\omega \in \Omega$,

$$\sup_{t_0 \leq t \leq a} |W_t^{(n)}(\omega)| \leq C(\omega) \quad \text{(for all } n \geq n_0(\omega))$$

and

$$\sup_{t_0 \le t \le a} |W_t^{(n)}(\omega) - W_t(\omega)| \to 0 \quad (n \to \infty) .$$

This is the case, for example, for the polygonal approximation at the points $t_0 < t_1 < \ldots < t_n = a$:

$$W_t^{(n)} = W_{t_k} + (W_{t_{k+1}} - W_{t_k}) \frac{t - t_k}{t_{k+1} - t_k}, \quad t_k \le t \le t_{k+1} ,$$

and

$$\delta_n = \max (t_{k+1} - t_k) \to 0.$$

Fig. 7:
Polygonal approxi-
mation of the sam-
ple functions of a
Wiener process.

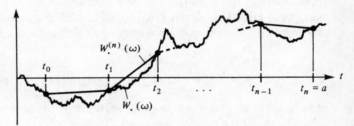

In the equation

$$X_t^{(n)} = c + \int_{t_0}^{t} f(s, X_s^{(n)}) \, ds + \int_{t_0}^{t} G(s, X_s^{(n)}) \, dW_s^{(n)}$$

the last integral is an ordinary Riemann-Stieltjes integral for the individual trajectories. Under certain conditions on the functions f and G, the sequence of the $X_t^{(n)}$ then converges with probability 1, uniformly on $[t_0, a]$, to the solution of the Stratonovich equation (10.3.3); that is,

$$\text{ac-}\lim_{n \to \infty} (\sup_{t_0 \le t \le a} |X_t^{(n)} - X_t|) = 0,$$

where X_t satisfies (10.3.3).

In this connection, we cite a result of Wong and Zakai [52] for the scalar case.

(10.3.4) Theorem. Suppose that $d = m = 1$ and that $\{W_t^{(n)}\}$ is a sequence of approximations of the Wiener process W_t with the properties mentioned above. Suppose that the functions $f(t, x)$ and $G(t, x)$ are continuous functions defined on $[t_0, T] \times R^1$ and that G has continuous partial derivatives G_t and G_x. Suppose that the functions f, G, and $G_x G$ satisfy a Lipschitz condition in x (see (6.2.4)). Suppose that

$$G(t, x) \ge \alpha > 0 \quad (\text{ or } \quad -G(t, x) \ge \alpha > 0)$$

and

$$|G_t(t, x)| \leqq \beta \, G(t, x)^2.$$

Suppose finally that the initial value c is independent of $W_t - W_{t_0}$ for $t \in [t_0, T]$ and that $X_t^{(n)}$ is the solution of the equation

$$X_t^{(n)} = c + \int_{t_0}^t f(s, X_s^{(n)}) \, ds + \int_{t_0}^t G(s, X_s^{(n)}) \, dW_s^{(n)}, \quad t_0 \leqq t \leqq T.$$

Then, for every finite $a \in [t_0, T]$,

$$\text{ac-}\lim_{n \to \infty} \left(\sup_{t_0 \leqq t \leqq a} |X_t^{(n)} - X_t| \right) = 0,$$

where X_t is the unique solution of the Stratonovich equation

(S) $dX_t = f(t, X_t) \, dt + G(t, X_t) \, dW_t$

or of the equivalent Itô equation

$$dX_t = \left(f(t, X_t) + \frac{1}{2} G_x(t, X_t) \, G(t, X_t) \right) dt + G(t, X_t) \, dW_t$$

with initial value $X_{t_0} = c$.

Thus, as a *rule of thumb*, one arrives at a Stratonovich equation when one shifts to a white noise or Wiener process in the result after evaluation of an ordinary integral.

Chapter 11
Stability of Stochastic Dynamic Systems

11.1 Stability of Deterministic Systems

Crudely speaking, stability of a dynamic system means insensitivity of the state of the system to small changes in the initial state or the parameters of the system. The trajectories of a stable system that are "close" to each other at a specific instant should therefore remain close to each other at all subsequent instants. For the theory of stability of deterministic systems, especially those systems that are described by ordinary differential equations, we refer to the monographs by Hahn [64] and Bhatia and Szegö [58]. We mention here only a few basic facts.

Suppose that

$$(11.1.1) \qquad \dot{X}_t = f(t, X_t), \quad X_{t_0} = c, \quad t \geq t_0,$$

is the ordinary differential equation for a d-dimensional state vector X_t. We assume that, for every initial value $c \in R^d$, there exists a global solution (that is, one defined on $[t_0, \infty))$ $X_t(c)$ (as is guaranteeed by the Lipschitz and boundedness conditions) and that $f(\cdot, x)$ is continuous. Furthermore, suppose that

$$f(t, 0) = 0 \text{ for all } t \geq t_0,$$

so that (11.1.1) has the solution $X_t \equiv 0$ corresponding to the initial condition $c = 0$. We shall refer to this solution as the **equilibrium position**. The equilibrium position is said to be **stable** if, for every $\varepsilon > 0$, there exists a $\delta = \delta(\varepsilon, t_0) > 0$ such that

$$\sup_{t_0 \leq t < \infty} |X_t(c)| \leq \varepsilon$$

whenever $|c| \leq \delta$. Otherwise, it is said to be **unstable**. The equilibrium position is said to be **asymptotically stable** if it is stable and if

$$\lim_{t \to \infty} X_t(c) = 0$$

for all c in some neighborhood of $x = 0$.

The definition of stability contains t_0 as a parameter. However, for a $\delta\,(\varepsilon, t_0) > 0$ corresponding to the definition, there exists a $\delta\,(\varepsilon, t_1) > 0$ for every $t_1 > t_0$.

The definition of stability just given leads to difficulties in that only in special cases can we solve equation (11.1.1) explicitly. As far back as 1892, A. M. Lyapunov developed a method (the so-called direct or second method) for determining stability without actually solving equation (11.1.1).

(11.1.2) Definition. A continuous scalar function $v\,(x)$ defined on a spherical neighborhood of the zero point

$$U_h = \{x : |x| \leqq h\} \subset R^d, \quad h > 0 \;,$$

is said to be **positive-definite** (in the sense of Lyapunov) if

$$v\,(0) = 0, \quad v\,(x) > 0 \quad (\text{for all } x \neq 0).$$

A continuous function $v\,(t, x)$ defined on $[t_0, \infty) \times U_h$ is said to be positive-definite if $v\,(t, 0) \equiv 0$ and there exists a positive-definite function $w\,(x)$ such that

$$v\,(t, x) \geqq w\,(x) \text{ for all } t \geqq t_0 \;.$$

A function v is said to be **negative-definite** if $-v$ is positive-definite. A continuous nonnegative function $v\,(t, x)$ is said to be **decrescent** (in Russian literature, it is said to have an arbitrarily small upper bound) if there exists a positive-definite function $u\,(x)$ such that

$$v\,(t, x) \leqq u\,(x) \text{ for all } t \geqq t_0.$$

It is said to be **radially unbounded** if

$$\inf_{t \geqq t_0} v\,(t, x) \longrightarrow \infty \quad (|x| \to \infty).$$

Every positive-definite function $v\,(x)$ that is independent of t is also decrescent.

If X_t is a solution of (11.1.1) and $v\,(t, x)$ is a positive-definite function with continuous first partial derivatives with respect to t and to the components x_i of x, then

$$V_t = v\,(t, X_t)$$

represents a function of t whose derivative is, by virtue of (11.1.1),

$$\dot{V}_t = \frac{\partial v}{\partial t} + \sum_{i=1}^{d} \frac{\partial v}{\partial x_i}\, f_i\,(t, X_t).$$

If $\dot{V}_t \leqq 0$, then X_t varies in such a way that values of V_t do not increase, that is, the "distance" of X_t from the equilibrium point measured by $v\,(t, X_t)$ does not increase. This elementary consideration leads to the following sufficient criteria discovered by Lyapunov.

(11.1.3) Theorem. a) If there exists a positive-definite function $v\,(t, x)$ with continuous first partial derivatives such that the derivative formed along the trajectories of

$$\dot{X}_t = f(t, X_t), \quad t \geq t_0, \quad f(t, 0) \equiv 0,$$

satisfies the inequality

$$\dot{v}(t, x) = \frac{\partial v}{\partial t} + \sum_{i=1}^{d} \frac{\partial v}{\partial x_i} f_i(t, x) \leq 0$$

in a half-cylinder

$$\{(t, x): t \geq t_0, |x| \leq h\}$$

then the equilibrium position of the differential equation is stable.

b) If there exists a positive-definite decrescent function $v(t, x)$ such that $\dot{v}(t, x)$ is negative-definite, then the equilibrium position is asymptotically stable.

A function $v(t, x)$ that satisfies the stability conditions of Theorem (11.1.3) is said to be a Lyapunov function corresponding to the differential equation in question.

(11.1.4) Example. The linear autonomous equation

$$\dot{X}_t = A X_t, \quad t \geq t_0, \quad X_{t_0} = c,$$

has the solution

$$X_t = e^{A(t-t_0)} c.$$

Let $\lambda_1, \dots, \lambda_d$ denote the eigenvalues of A. Then, the equilibrium position is asymptotically stable if and only if

$$(11.1.5) \quad \text{Re}(\lambda_i) < 0, \quad i = 1, \dots, d.$$

If at least one eigenvalue has a positive real part, it is unstable. If some of the real parts vanish, the equilibrium position is stable (though not asymptotically stable) provided the elementary divisors corresponding to the eigenvalues with vanishing real parts are all simple. If any of the elementary divisors are of higher order, the equilibrium position is unstable. One can see whether (11.1.5) holds or not by using the criteria of Routh, Hurwitz, and others (see Hahn [64]) or one can see whether the equation

$$(11.1.6) \quad A'P + PA = -Q,$$

has, for some positive-definite Q, a positive-definite matrix P as its solution. Then, we can choose the Lyapunov function

$$v(x) = x'Px > 0$$

for which

$$\dot{v}(x) = 2(Px)'Ax = x'PAx + x'A'Px = -x'Qx < 0.$$

11.2 The Basic Ideas of Stochastic Stability Theory

When we try to carry over the principles of deterministic Lyapunov stability theory to the stochastic case, we encounter two problems:

1. What is a suitable definition of stability?

2. What is the corresponding definition of a Lyapunov function? With what must we replace the inequality $\dot{v} \leq 0$ in order to get assertions of the type of Theorem (11.1.3)?

Of the numerous attempts along these lines in recent years, the one that has received general acceptance is that of Bucy [59], who recognized in 1965 that a stochastic Lyapunov function should have the supermartingale property, which corresponds to a quite strict stability concept. For a detailed account and further references, we refer the reader to the books by Kushner [72], Bucy and Joseph [61], Morozan [73], and, above all, the profound work by Khas'minskiy [65], which we shall essentially follow in the present book.

In all that follows, we shall make the

(11.2.1) **Assumptions.** Suppose that

$$(11.2.2) \qquad dX_t = f(t, X_t)\,dt + G(t, X_t)\,dW_t, \quad X_{t_0} = c, \quad t \geq t_0,$$

is a stochastic differential equation that satisfies the assumptions of the existence-and-uniqueness theorem (6.2.1) and has continuous coefficients with respect to t. Then, in accordance with Theorem (9.3.1), corresponding to every c that is independent of W, there exists a unique global solution $X_t = X_t(c)$ on $[t_0, \infty)$, which represents a d-dimensional diffusion process with drift vector $f(t, x)$ and diffusion matrix $B(t, x) = G(t, x)\, G(t, x)'$. Furthermore, let us assume once and for all that c is with probability 1 a *constant*. Then, Theorem (7.1.2) implies the existence of all moments $E\,|X_t|^k$ for $k > 0$ and also

$$P(X_t \in B\,|\,X_{t_0} = c) = P\,[X_t(c) \in B].$$

The solution beginning at the instant $s \geq t_0$ at the point x will be denoted by $X_t(s, x)$. Finally,

$$f(t, 0) = 0, \quad G(t, 0) = 0 \text{ for all } t \geq t_0,$$

so that the equilibrium position $X_t(0) \equiv 0$ is the unique solution of the differential equation with initial value $c = 0$. The case in which the equilibrium position is a solution of the undisturbed equation and the disturbance acts even at $x = 0$ (that is, $f(t, 0) \equiv 0$ but $G(t, 0) \not\equiv 0$), is considered in remark (11.2.18).

To clarify the basic idea of stochastic stability theory, we make first of all the following observation:

Let X_t denote the solution of equation (11.2.2) and let $v(t, x)$ denote a positive-definite function defined everywhere on $[t_0, \infty) \times R^d$ that has continuous partial derivatives v_t, v_{x_i} and $v_{x_i x_j}$. Then, the process

$$V_t = v(t, X_t)$$

has, in accordance with Itô's theorem, a stochastic differential. If we denote by L the extension of the infinitesimal operator A of X_t to all functions that are continuously differentiable with respect to t and twice continuously differentiable with respect to the x_i (see section 9.4), that is,

(11.2.3)

$$L = \frac{\partial}{\partial t} + \mathfrak{D}, \quad L \supset A,$$

$$\mathfrak{D} = \sum_{i=1}^{d} f_i(t, x) \frac{\partial}{\partial x_i} + \frac{1}{2} \sum_{i=1}^{d} \sum_{j=1}^{d} (G(t, x) G(t, x)')_{ij} \frac{\partial^2}{\partial x_i \partial x_j},$$

then, in accordance with formula (5.3.9a),

(11.2.4) $dV_t = (L v(t, X_t)) dt + \sum_{i=1}^{d} \sum_{j=1}^{m} v_{x_i}(t, X_t) G_{ij}(t, X_t) dW_t^j.$

Now, a stable system should have the property that V_t does not increase, that is, $dV_t \leqq 0$. This would mean that the ordinary definition of stability holds for each single trajectory $X_.(\omega)$. However, because of the presence of the fluctuational term in (11.2.4), the condition $dV_t \leqq 0$ can be satisfied only in degenerate cases. Therefore, it makes sense to require instead that X_t not run "up hill" on the average, that is,

$$E(dV_t) \leqq 0.$$

Since

$$E(dV_t) = E(L v(t, X_t) dt)$$

this requirement will be satisfied if

(11.2.5) $L v(t, x) \leqq 0$ for all $t \geqq t_0$, $x \in R^d$.

This is the stochastic analogue of the requirement that $\dot{v} \leqq 0$ in the deterministic case and it reduces to that case if G vanishes. We shall refer to the function $v(t, x)$ used here as a **Lyapunov function** corresponding to the stochastic differential equation (11.2.2).

For what stability concept is (11.2.5) a sufficient condition? In this connection, we remember that, in accordance with Theorem (5.1.1b), the second integral in

$$V_t = v(t_0, c) + \int_{t_0}^{t} L v(s, X_s) ds + \int_{t_0}^{t} \sum_{i=1}^{d} \sum_{j=1}^{m} v_{x_i}(s, X_s) G_{ij}(s, X_s) dW_s^j$$

is a martingale. Here, the accompanying family $\mathfrak{F}_t$ of sigma-algebras is the one defined in (6.1.2). Therefore, for $t \geqq s$, we have by virtue of (11.2.5)

$$E\left(V_t - V_s \mid \mathfrak{F}_s\right) = E\left(\int_s^t L\,v\left(u, X_u\right)\,du \;\middle|\; \mathfrak{F}_s\right) \leqq 0$$

or

$$E\left(V_t \mid \mathfrak{F}_s\right) \leqq V_s,$$

that is, *under condition (11.2.5), V_t is a (positive) supermartingale.* For every interval $[a, b] \subset [t_0, \infty)$, the supermartingale inequality yields

$$P\left[\sup_{a \leqq t \leqq b} v\left(t, X_t\right) \geqq \varepsilon\right] \leqq \frac{1}{\varepsilon} E\,v\left(a, X_a\right)$$

and, in particular, for $a = t_0$ and $b \to \infty$ (the initial value c is constant!),

$$P\left[\sup_{t_0 \leqq t < \infty} v\left(t, X_t\right) \geqq \varepsilon\right] \leqq \frac{1}{\varepsilon} v\left(t_0, c\right), \text{for all } \varepsilon > 0, \quad c \in R^d.$$

Now, $v\left(t, x\right)$ is positive-definite. To simplify the analysis, we exclude Lyapunov functions that become arbitrarily small for large $|x|$ by requiring that $v\left(t, x\right) \geqq w\left(x\right) \to \infty$ as $|x| \to \infty$, that is, by requiring that v be radially unbounded. Then, for arbitrary positive ε_1 and ε_2, there exists an $\varepsilon = \varepsilon\left(\varepsilon_1\right)$ and (subsequently) a $\delta = \delta\left(\varepsilon_1, \varepsilon_2, t_0\right)$ such that

$$v\left(t, x\right) \geqq \varepsilon, \quad \text{if } |x| \geqq \varepsilon_1,$$

and

$$v\left(t_0, c\right)/\varepsilon \leqq \varepsilon_2, \quad \text{if } |c| \leqq \delta,$$

so that, with $X_t = X_t\left(c\right)$,

$$P\left[\sup_{t_0 \leqq t < \infty} |X_t\left(c\right)| \geqq \varepsilon_1\right] \leqq \varepsilon_2 \text{ for } |c| \leqq \delta.$$

Therefore, for arbitrary positive ε_1 and ε_2, there exists a positive δ such that all the trajectories with starting point at c, where $|c| \leqq \delta$, range, for all time, in an ε_1-neighborhood of the zero point—with the exception of a set of trajectories that occur with some probability not exceeding ε_2. This is the $\varepsilon\delta$-formulation of the following stability definition:

(11.2.6) **Definition.** Suppose that the assumptions (11.2.1) are satisfied. Then the equilibrium position is said to be **stochastically stable** ("stable in probability" in Khas'minskiy's terminology [65] or "stable with probability 1" in Kushner's terminology [72]) if, for every $\varepsilon > 0$,

$$\lim_{c \to 0} P\left[\sup_{t_0 \leqq t < \infty} |X_t\left(c\right)| \geqq \varepsilon\right] = 0$$

Otherwise, it is said to be **stochastically unstable.** The equilibrium position is said to be **stochastically asymptotically stable** if it is stochastically stable and

$$\lim_{c \to 0} P \left[\lim_{t \to \infty} X_t(c) = 0 \right] = 1.$$

The equilibrium position is said to be stochastically asymptotically stable **in the large** if it is stochastically stable and

$$P \left[\lim_{t \to \infty} X_t(c) = 0 \right] = 1$$

for *all* $c \in R^d$.

It should be noted that, in the case $G(t, x) \equiv 0$, these definitions reduce to the corresponding deterministic definitions.

Thus, we have shown that the existence of a positive-definite function $v(t, x)$ with the property (11.2.5) implies stochastic stability.

For asymptotic stability, we must exclude the case $L v(t, x) = 0$ (because then there is no average trend in direction $x = 0$) and replace (11.2.5) with the more stringent condition

(11.2.7) $L v(t, x)$ is negative-definite.

Suppose also that v is decrescent and, together with $-L v$, radially unbounded. Now, V_t is, by virtue of (11.2.7), again a positive supermartingale. Therefore, in accordance with section 1.9, there exists

$$\text{ac-} \lim_{t \to \infty} V_t = V^{(c)} \geq 0.$$

Here, the limit $V^{(c)}$ may depend on the initial point c. If $V^{(c)}$ were at least equal to some positive c_1 on an ω-set B_c with positive probability, we would have, for these ω,

$$v(t, X_t) \geq c_2 > 0 \text{ for all } t \geq \tau(\omega),$$

and, by virtue of the decrescence of v,

$$|X_t| \geq c_3 > 0 \text{ for all } t \geq \tau(\omega).$$

The assumption (11.2.7) and the radial unboundedness of $-L v$ implies the existence of a positive c_4 such that

$$L v(t, x) \leq -c_4 \quad \text{for } |x| \geq c_3.$$

By virtue of (11.2.4),

$$0 \leq V_t = v(t_0, c) + \int_{t_0}^{t} L v(s, X_s) \, ds + M_t,$$

where M_t is a scalar martingale such that $E M_t = 0$ and, by virtue of (11.2.7), $M_t \geq -v(t_0, c)$. The supermartingale inequality yields

$$P\left[\sup_{t \geq t_0} M_t \geq \varepsilon\right] \leq \frac{v(t_0, c)}{\varepsilon}.$$

For all $\omega \in B_c \cap [\sup M_t < \varepsilon]$, we have, for $t \geq \tau(\omega)$,

$$0 \leq V_t \leq v(t_0, c) - c_4(t - \tau(\omega)) + \varepsilon.$$

When we let t approach ∞, this leads to a contradiction; that is, $B_c \cap [\sup M_t < \varepsilon]$ has probability 0. Therefore,

$$P(B_c) = P(B_c \cap [\sup M_t \geq \varepsilon]) \leq \frac{v(t_0, c)}{\varepsilon}$$

and, finally,

$$P\left[\lim_{t \to \infty} v(t, X_t) = 0\right] \geq 1 - \frac{v(t_0, c)}{\varepsilon}.$$

By virtue of the radial unboundedness of v, we therefore have

$$P\left[\lim_{t \to \infty} X_t(c) = 0\right] \geq 1 - \frac{v(t_0, c)}{\varepsilon} \to 1 \quad (c \to 0).$$

Thus, the equilibrium position is stochastically asymptotically stable.

This proves a special case of the following general theorem (see Khas'minskiy [65]):

(11.2.8) **Theorem.** Suppose that the assumptions (11.2.1) are satisfied.
a) Suppose that there exists a positive-definite function $v(t, x)$ defined on a half-cylinder $[t_0, \infty) \times U_h$, $U_h = \{x : |x| < h\}$, where $h > 0$, that is everywhere, with the possible exception of the point $x = 0$, continuously differentiable with respect to t and twice continuously differentiable with respect to the components x_i of x. Furthermore,

$$L v(t, x) \leq 0, \quad t \geq t_0, \quad 0 < |x| \leq h,$$

where

$$L = \frac{\partial}{\partial t} + \sum_{i=1}^{d} f_i(t, x) \frac{\partial}{\partial x_i} + \frac{1}{2} \sum_{i=1}^{d} \sum_{j=1}^{d} (G(t, x) G(t, x)')_{ij} \frac{\partial^2}{\partial x_i \partial x_j}.$$

Then, the equilibrium position of equation (11.2.2) is stochastically stable.

b) If, in addition, $v(t, x)$ is decrescent and $L v(t, x)$ is negative-definite, then the equilibrium position is stochastically asymptotically stable. In both cases,

$$P\left[\sup_{t \geq s} v(t, X_t(s, x)) \geq \varepsilon\right] \leq \frac{v(s, x)}{\varepsilon}, \quad \varepsilon > 0, \quad s \geq t_0.$$

c) If the assumptions of part b) hold for a radially unbounded function $v(t, x)$ defined everywhere on $[t_0, \infty) \times R^d$, then the equilibrium position is stochastically asymptotically stable in the large.

(11.2.9) Remark. A sufficient condition for negative-definiteness of $L\,v$ is the existence of a constant $k > 0$ such that

$$L\,v\,(t, x) \leqq -k\,v\,(t, x).$$

(11.2.10) Remark. For an *autonomous* equation

$$dX_t = f\,(X_t)\,dt + G\,(X_t)\,dW_t, \quad f\,(0) = 0, \quad G\,(0) = 0,$$

it is sufficient to consider a function $v\,(t, x) \equiv v\,(x)$ that is independent of t. Khas'-minskiy [65] has shown that the existence of a Lyapunov function $v\,(x)$ is a necessary condition for stochastic stability as long as the disturbance is "nondegenerate", that is, as long as

$$(11.2.11) \quad y\,G\,(x)\,G\,(x)'\,y' \geqq m\,(x)\,|y|^2 \text{ for all } y \in R^d, \quad x \in U_h,$$

where $m\,(x)$ is positive-definite. Under condition $(11.2.11)$, stochastic stability implies stochastic asymptotic stability (see Khas'minskiy [65], p. 213).

We state the following criterion for stochastic instability (Khas'minskiy [65]):

(11.2.12) Theorem. Suppose that the assumptions $(11.2.1)$ are satisfied. Suppose that there exists a function $v\,(t, x)$ defined on $[t_0, \infty) \times \{0 < |x| \leqq h\}$ that has continuous partial derivatives with respect to t and continuous second partial derivatives with respect to the x_i. Then, if

$$\lim_{x \to 0} \inf_{t \geqq t_0} v\,(t, x) = \infty$$

and

$$(11.2.13) \quad \sup_{t \geqq t_0,\, \varepsilon < |x| \leqq h} L\,v\,(t, x) < 0 \text{ for all } 0 < \varepsilon < h,$$

or, instead of $(11.2.13)$, only $L\,v\,(t, x) \leqq 0$ (for $x \neq 0$) and, in addition,

$$(11.2.14) \quad \begin{array}{l} y\,G\,(t, x)\,G\,(t, x)'\,y' \geqq m\,(x)\,|y|^2 \text{ for all } y \in R^d, \quad |x| \leqq h, \\ t \geqq t_0, \quad m \text{ positive-definite,} \end{array}$$

then the equilibrium position of $(11.2.1)$ is stochastically unstable. Furthermore,

$$P\,[\sup_{t \geqq t_0} |X_t\,(c)| < h] = 0 \text{ for all } c \in U_h.$$

(11.2.15) Example. The solution of the scalar linear autonomous homogeneous differential equation $(d = m = 1)$

$$(11.2.16) \quad dX_t = A\,X_t\,dt + B\,X_t\,dW_t, \quad X_{t_0} = c,$$

is, in accordance with Corollary $(8.4.3b)$,

$$X_t\,(c) = c\,\exp\,((A - B^2/2)\,(t - t_0) + B\,(W_t - W_{t_0})).$$

From this we conclude the following: a) for no positive ε does there exist a positive δ such that *all* the sample functions of the bundle of sample functions originating at $c \neq 0, |c| < \delta$ remain in an ε-neighborhood of $x = 0$.

b) Since

$$\text{ac-}\lim_{t \to \infty} \frac{W_t - W_{t_0}}{t - t_0} = 0$$

the equilibrium position is stochastically asymptotically stable in the large for

$$A < B^2/2$$

and stochastically unstable for

$$A \geq B^2/2 > 0.$$

Let us derive the same result with the aid of Lyapunov's technique. For X_t, the operator L takes the form

$$L = \frac{\partial}{\partial t} + A x \frac{\partial}{\partial x} + \frac{1}{2} B^2 x^2 \frac{\partial^2}{\partial x^2} .$$

A trial with $v(t, x) \equiv v(x) = |x|^r$, where $r > 0$, leads, for $x \neq 0$, to

$$L(|x|^r) = \left(A + \frac{1}{2} B^2 (r-1)\right) r |x|^r.$$

As long as $A < B^2/2$, we can choose r such that $0 < r < 1 - 2 A/B^2$ and hence satisfy the condition

$$L v \leq -k v$$

which, according to remark (11.2.9), is sufficient for stochastic asymptotic stability. Since $v(x) = |x|^r$ is radially unbounded, we have, in accordance with Theorem (11.2.8c), stochastic asymptotic stability in the large. As an estimate, we have

$$P[\sup_{t_0 \leq t < \infty} |X_t(c)| \geq \varepsilon] \leq \frac{|c|^r}{\varepsilon^r}.$$

If $A \geq B^2/2 > 0$, we choose $v(x) = -\log |x|$ and we obtain $L v = -A + B^2/2 \leq 0$. Since condition (11.2.14) is satisfied for $B \neq 0$ with $m(x) = B^2 |x|^2$, stochastic instability is ensured by Theorem (11.2.12). For $B = 0$, the system is deterministic and we see from example (11.1.4) that we have asymptotic stability for $A < 0$, simple stability for $A = 0$, and instability for $A > 0$.

(11.2.17) **Remark.** In example (11.2.15), the "undisturbed" system ($B = 0$) is therefore stable only for $A \leq 0$, and the addition of a "disturbance term" does not change this property. However, the system can be stabilized even in the case of $A > 0$ by adding a disturbance term of strength (= diffusion coefficient) $B^2 x^2$ provided we choose $B^2/2 > A$.

Now, equation (11.2.16) is an abstraction of

$$\dot{X}_t = (A + B\,\eta_t)\,X_t,$$

where η_t is a Gaussian stationary process with $E\,\eta_t = 0$ and a delta-like covariance function. If we shift from η_t to the white noise ξ_t only after we have found the solution

$$X_t = c\,e^{A(t-t_0)+B\int_{t_0}^{t}\eta_s\,ds},$$

we obtain

$$X_t = c\,e^{A(t-t_0)+B(W_t-W_{t_0})}.$$

In accordance with section (10.2), this is the solution of equation (11.2.16), now interpreted as the Stratonovich equation, or of the equivalent Itô equation

$$dX_t = (A + B^2/2)\,X_t\,dt + B\,X_t\,dW_t,\quad X_{t_0} = c,$$

whose equilibrium position is now stochastically asymptotically stable for arbitrary $B \neq 0$ if $A < 0$ and stochastically unstable if $A \geq 0$.

With either interpretation, an asymptotically stable undisturbed system ($A < 0$, $B = 0$) remains stochastically asymptotically stable upon addition of arbitrarily strong disturbances (only ordinary stability $A = 0$ is destroyed). This property disappears with Itô equations for $d \geq 3$ and with Stratonovich equations for $d \geq 2$ (see Khas'minskiy [65], p. 222).

(11.2.18) **Remark.** The use of Lyapunov's method depends on knowledge of Lyapunov functions $v\,(t,\,x)$. As in the deterministic case, there are a number of techniques that can be used to find suitable functions. For example, one can seek a positive-definite solution of the equation $L\,v = 0$ or of the inequality $L\,v \leq 0$ (see Kushner [72], pp. 55-76, for a number of examples). The choice

$$v\,(x) = x'\,C\,x,$$

where C is a positive-definite matrix, leads to the goal if

$$L\,v = 2\,f\,(t,\,x)'\,C\,x + \operatorname{tr}\,(G\,(t,\,x)\,G\,(t,\,x)'\,C) \leq 0$$

in some neighborhood of $x = 0$ for $t \geq t_0$.

(11.2.19) **Remark.** If the functions f and G in equation (11.2.2) are such that $f\,(t,\,0) \equiv 0$ but $G\,(t,\,0) \not\equiv 0$, the equilibrium position is a solution of the undisturbed system but no longer of the disturbed system. However, we can in principle apply our definitions of stability (11.2.6). Suppose, that for arbitrary d and m,

$$dX_t = A\,X_t\,dt + B\,(t)\,dW_t,\quad X_{t_0} = c,$$

where the undisturbed system is asymptotically stable (see example (11.1.4)). According to Corollary (8.2.4), the solution of this equation is

$$X_t = e^{A(t-t_0)} c + \int_{t_0}^{t} e^{A(t-s)} B(s) \, dW_s.$$

By assumption $e^{A(t-t_0)} c \to 0$ as $t \to \infty$ for all $c \in R^d$. The second term is normally distributed with mean 0 and

$$E \left| \int_{t_0}^{t} e^{A(t-s)} B(s) \, dW_s \right|^2 = \int_{t_0}^{t} |e^{A(t-s)} B(s)|^2 \, ds$$

(see remark (8.2.9)).

Let λ_i denote the eigenvalues of A and let us set

$$-r = \max_{i} (\text{Re} (\lambda_i)) < 0.$$

Then,

$$|e^{A(t-s)} B(s)|^2 \leq |B(s)|^2 e^{-2r(t-s)},$$

so that

$$\int_{t_0}^{t} |e^{A(t-s)} B(s)|^2 \, ds \leq \int_{t_0}^{t} e^{-2r(t-s)} |B(s)|^2 \, ds$$

$$\leq e^{-rt} \int_{t_0}^{t/2} |B(s)|^2 \, ds + \int_{t/2}^{t} |B(s)|^2 \, ds$$

$$\to 0 \ (t \to \infty),$$

this last holding under the condition

$$(11.2.20) \qquad \int_{t_0}^{\infty} |B(s)|^2 \, ds < \infty.$$

This means that

$$\text{qm-}\lim_{t \to \infty} X_t = 0.$$

Moreover, by virtue of a theorem of Khas'minskiy ([65, pp. 307-310), for every initial value c, we have under the condition (11.2.20)

$$P [\lim_{t \to \infty} X_t = 0] = 1.$$

11.3 Stability of the Moments

The definition of stochastic stability involves the overall behavior of the sample functions $X_. (\omega)$ on the interval $[t_0, \infty)$. For given $\varepsilon > 0$ and starting point c, we group together all those sample functions that remain for all t in an ε-neighborhood of the zero point $x = 0$. The probability of this set must in the case of a stable system be close to 1 provided $|c|$ is small.

In many cases, the following concept (historically, the older one) is more feasible: For fixed t, we take the average over all possible values of X_t or of a function $g(X_t)$; that is, we take $E\, g(X_t) = g_1(t)$ and examine the so-obtained deterministic function g_1 for its behavior in the interval $[t_0, \infty)$. We shall treat the cases $g(x) = x$, $g(x) = |x|^p$ (for $p > 0$), and $g(x) = x\, x'$.

(11.3.1) **Definition.** Suppose that the stochastic differential equation

$$(11.3.2) \qquad dX_t = f(t, X_t)\, dt + G(t, X_t)\, dW_t, \quad X_{t_0} = c \text{ (fixed)},$$

satisfies the assumptions (11.2.1). The equilibrium position is said to be **stable in pth mean** (where $p > 0$) if, for every $\varepsilon > 0$, there exists a $\delta > 0$ such that

$$\sup_{t_0 \leq t < \infty} E\, |X_t(c)|^p \leq \varepsilon \text{ for } |c| \leq \delta.$$

The equilibrium position is said to be **asymptotically stable in pth mean** if it is stable in pth mean and if

$$\lim_{t \to \infty} E\, |X_t(c)|^p = 0 \quad \text{for all } c \text{ in a neighborhood of } x = 0.$$

It is said to be **exponentially stable in pth mean** if there exist positive constants c_1 and c_2 such that, for all sufficiently small c,

$$E\, |X_t(c)|^p \leq c_1 \, |c|^p \, e^{-c_2 (t - t_0)}.$$

In the case $p = 1$ or $p = 2$, we speak of (asymptotic or exponential) **stability in mean** or **in mean square** respectively. The equilibrium position is said to possess a **stable expectation value** $m_t = E\, X_t(c)$ or a **stable second moment** $P(t) = E\, X_t(c)\, X_t(c)'$ if, for every $\varepsilon > 0$, there exists a $\delta > 0$ such that

$$\sup_{t_0 \leq t < \infty} |E\, X_t(c)| \leq \varepsilon \text{ for all } |c| \leq \delta,$$

or

$$\sup_{t_0 \leq t < \infty} |E\, X_t(c)\, X_t(c)'| \leq \varepsilon \text{ for all } |c| \leq \delta$$

respectively.

(11.3.3) **Remark.** a) Since the function $(E\, |X|^p)^{1/p}$ is monotonically increasing when $p > 0$, (asymptotic, exponential) stability in pth mean implies the same stability in qth mean (for all $0 < q < p$). In particular, stability in mean square implies stability in mean.

b) It follows from the Chebychev inequality that

$$\sup_{t_0 \leq t < \infty} P\left[|X_t(c)| \geq \varepsilon\right] \leq \frac{1}{\varepsilon^p} \sup_{t_0 \leq t < \infty} E\,|X_t(c)|^p,$$

that is, stability in pth mean implies

$$\lim_{c \to 0}\ \sup_{t_0 \leq t < \infty} P\left[|X_t(c)| \geq \varepsilon\right] = 0 \text{ for all } \varepsilon > 0.$$

This is called **weak stochastic stability**. It is obviously weaker than stochastic stability

$$\lim_{c \to 0} P\left[\sup_{t_0 \leq t < \infty} |X_t(c)| \geq \varepsilon\right] = 0 \text{ for all } \varepsilon > 0.$$

For an autonomous linear equation, the two concepts are, however, equivalent according to Khas'minskiy ([65], p. 296).

c) Stability in mean square is equivalent to stability of the second moment. This is true because, on the one hand, for $P(t) = E\,X_t(c)\,X_t(c)' = (E\,X_t^i\,X_t^j)$

$$|P(t)| \leq E\,|X_t(c)|^2,$$

and, on the other hand,

$$E\,|X_t(c)|^2 = \operatorname{tr} P(t) = \sum_{i=1}^{d} P_{ii}(t).$$

d) Since $|E\,X| \leq E\,|X|$, stability in mean implies stability of the expectation value $E\,X_t = m_t$. However, the converse does not in general hold! Therefore, stability of the expectation value is a necessary condition for stability in mean square.

e) For the covariance matrix $K(t) = E\,(X_t - E\,X_t)\,(X_t - E\,X_t)'$, we have

$$K(t) = P(t) - m_t\,m_t'.$$

From a combination of c) and d), we see that stability in mean square implies stability of $K(t)$. This last is identical to stability of $E\,|X_t - m_t|^2$.

(11.3.4) **Example.** We have already, in example (11.2.15), investigated the stability behavior of the scalar linear autonomous homogeneous differential equation

$$dX_t = A\,X_t\,dt + B\,X_t\,dW_t, \quad X_{t_0} = c,$$

For the moments, we obtain from example (8.4.8)

$$E\,|X_t(c)|^p = |c|^p\,e^{(p(A - B^2/2) + p^2 B^2/2)(t - t_0)}.$$

From this we see that the equilibrium position is exponentially stable in pth mean if and only if

$$A < (1 - p) \, B^2/2 \, .$$

Thus, the points of stochastic asymptotic stability of the region $A < B^2/2$ are just those for which exponential stability in pth mean exists for all sufficiently small $p > 0$. This holds true in general for autonomous linear equations (see remark (11.4.17)).

Conversely, from the fact that the second moment $E \, |X_t|^2$ approaches 0 exponentially, we conclude that the equilibrium position is stochastically stable (see Kozin [70], Gikhman and Skorokhod [36], p. 331). We cite more generally (Khas'minskiy [65], pp. 232-237).

(11.3.5) **Theorem.** A sufficient condition for exponential stability in pth mean is the existence of a function $v \, (t, \, x)$, that is defined and continuous on $[t_0, \, \infty) \times R^d$, that for all $x \neq 0$ is once continuously differentiable with respect to t and twice continuously differentiable with respect to the x_i, and that satisfies the inequalities

$$c_1 \, |x|^p \leqq v \, (t, \, x) \leqq c_2 \, |x|^p$$

and

$$L \, v \, (t, \, x) \leqq - c_3 \, |x|^p$$

for certain positive constants c_1, c_2, and c_3.

Then, there exists a positive constant c_4 and an almost certainly finite random variable $K \, (c)$ dependent on $c \in R^d$ such that

$$|X_t \, (c)| \leqq K \, (c) \, e^{- c_4 \, (t - t_0)} \text{ for all } t \geqq t_0,$$

for almost all sample functions starting at c .

11.4 Linear Equations

For the linear homogeneous equation (see section 8.5)

$$(11.4.1) \quad dX_t = A \, (t) \, X_t \, dt + \sum_{i=1}^{m} B_i \, (t) \, X_t \, dW_t^i, \quad X_{t_0} = c, \quad t \geqq t_0,$$

considerably more precise stability assertions can be made for the equilibrium position $X_t \equiv 0$. In accordance with our assumptions (11.2.1), the $d \times d$ matrices $A \, (t), B_1 \, (t), \dots, B_m \, (t)$ are continuous functions on the interval $t \geqq t_0$. Suppose again that $c \in R^d$ is a constant.

Stability of the first and second moments acquires for (11.4.1) considerable significance in that simple ordinary differential equations can now be written for them. In accordance with Theorem (8.5.5), we have

a) $E X_t = m_t$ is the unique solution of the equation

(11.4.2) $\quad \dot{m}_t = A(t) m_t, \quad m_{t_0} = c.$

b) $E X_t X_t' = P(t)$ is the unique nonnegative-definite symmetric solution of the equation

$$(11.4.3) \quad \dot{P}(t) = A(t) P(t) + P(t) A(t)' + \sum_{i=1}^{m} B_i(t) P(t) B_i(t)', \quad P(t_0) = E c c'.$$

Therefore, stability of the expectation value is, for a linear equation, equivalent to stability of the undisturbed deterministic equation (11.4.2). Also, stability of the second moments of X_t (which, in accordance with section 11.3, is identical to stability in mean square) is equivalent to stability of the deterministic matrix equation (11.4.3).

Since

$$P_{ij}(t) = E X_t^i X_t^j = E X_t^j X_t^i = P_{ji}(t)$$

so that $P(t)$ is symmetric, (11.4.3) represents a system of $d(d+1)/2$ linear equations.

Let us group the $d(d+1)/2$ elements $P_{ij}(t)$, where $i \geq j$, in such a way as to form a vector $\mathfrak{p}(t)$. Then, (11.4.3) can be written in the form

(11.4.4) $\quad \dot{\mathfrak{p}}(t) = \mathfrak{A}(t) \mathfrak{p}(t), \quad \mathfrak{p}(t_0) = E \mathfrak{c},$

where $\mathfrak{c}$ is the $[d(d+1)/2]$-dimensional vector corresponding to the matrix $c c'$. Therefore, mean-square stability of (11.4.1) is identical to ordinary stability of the equilibrium position $\mathfrak{p} \equiv 0$ of the linear equation (11.4.4). This last condition is especially easy to check if (11.4.1) is autonomous, so that (11.4.4) becomes a system with constant coefficients. The reader should compare the criteria of Theorem (11.4.11) and Corollary (11.4.14). We recall that exponential stability of (11.4.4) implies stochastic stability of (11.4.1).

(11.4.5) **Example.** The second-order scalar differential equation with constant but "noisy" coefficients

$$\ddot{Y}_t + (b_0 + b\, \xi_t^1)\, \dot{Y}_t + (a_0 + a\, \xi_t^2)\, Y_t = 0, \quad t \geq 0,$$

where ξ_t^1 and ξ_t^2 are uncorrelated scalar white noise processes, is equivalent to the linear stochastic differential equation

$$dX_t = \begin{pmatrix} 0 & 1 \\ -a_0 & -b_0 \end{pmatrix} X_t\, dt + \begin{pmatrix} 0 & 0 \\ 0 & -b \end{pmatrix} X_t\, dW_t^1 + \begin{pmatrix} 0 & 0 \\ -a & 0 \end{pmatrix} X_t\, dW_t^2.$$

Here, we have again set

$$X_t = \begin{pmatrix} Y_t \\ \dot{Y}_t \end{pmatrix}$$

(see example (8.1.4)). The differential equation (11.4.2) for the expectation value m_t is

$$\dot{m}_t = \begin{pmatrix} 0 & 1 \\ -a_0 & -b_0 \end{pmatrix} m_t$$

and it is asymptotically stable if and only if both a_0 and b_0 are positive. Equation (11.4.3) for the 2×2 matrix $P(t)$ of the second moments yields

$$\dot{P}_{11} = 2 P_{12},$$
$$\dot{P}_{12} = -a_0 P_{11} - b_0 P_{12} + P_{22},$$
$$\dot{P}_{22} = a^2 P_{11} - 2 a_0 P_{12} + (b^2 - 2 b_0) P_{22}.$$

The 3×3 matrix $\mathfrak{A}$ in equation (11.4.4) is

$$\mathfrak{A} = \begin{pmatrix} 0 & 2 & 0 \\ -a_0 & -b_0 & 1 \\ a^2 & -2 a_0 & b^2 - 2 b_0 \end{pmatrix},$$

and its characteristic equation is

$$-\det(\mathfrak{A} - \lambda I) = \lambda^3 + \lambda^2 (3 b_0 - b^2) + \lambda (4 a_0 + 2 b_0^2 - b_0 b^2)$$
$$+ 2 (2 a_0 b_0 - a_0 b^2 - a^2) = 0.$$

The real parts of its roots are then negative if and only if

$$b^2 < 2 b_0, \quad a^2 < (2 b_0 - b^2) a_0$$

(by the Routh-Hurwitz criterion). Therefore, for fixed a_0 and b_0, the intensity of the disturbance must not exceed a certain value if the (exponential) stability in mean square is not to be destroyed. For the nth-order scalar differential equation with disturbed constant coefficients, compare Theorem (11.5.2).

(11.4.6) **Example.** Gikhman ([36] and [63], pp. 320-328) has investigated second-order scalar equations of the form

(11.4.7) $$\ddot{Y}_t + (a(t) + b(t) \eta_t) Y_t = 0, \quad t \geq t_0,$$

where η_t can be a general disturbance process (the derivative of a martingale). The stability properties of Y_t are closely connected with those of the undisturbed equation

$$\ddot{z}(t) + a(t) z(t) = 0.$$

For example, if η_t is a white noise, if, for every solution $z(t)$ of the last equation, $z(t) \to 0$ and

$$\int_{t_0}^{\infty} z(t)^2 \, dt < \infty,$$

and if $b(t)$ is bounded, then $E Y_t^2 \to 0$ uniformly for all initial values such that $|Y_{t_0}| + |\dot{Y}_{t_0}| \leq R$ as $t \to \infty$. Furthermore, the equilibrium position of equation (11.4.7) is then stochastically stable.

(11.4.8) **Example.** The general homogeneous equation (11.4.1) for the case $d = m = 1$

$$dX_t = A(t) X_t \, dt + B(t) X_t \, dW_t, \quad X_{t_0} = c,$$

has, by Corollary (8.4.3), the solution

$$X_t = c \exp \left(\int_{t_0}^{t} (A(s) - B(s)^2/2) \, ds + \int_{t_0}^{t} B(s) \, dW_s \right).$$

From Theorem (8.4.5), we have

$$E X_t = c \exp \left(\int_{t_0}^{t} A(s) \, ds \right)$$

and

$$E |X_t|^p = |c|^p \exp \left(p \int_{t_0}^{t} A(s) \, ds + \frac{p(p-1)}{2} \int_{t_0}^{t} B(s)^2 \, ds \right).$$

From all this, we conclude that the equilibrium position is asymptotically stable (resp. stable, resp. unstable) in pth mean if and only if

$$\lim_{t \to \infty} \sup \int_{t_0}^{t} \left(p A(s) + \frac{p(p-1)}{2} B(s)^2 \right) ds$$

is $-\infty$ (resp. is less than ∞, resp. is ∞). In particular, this yields criteria for the first and second moments. Similarly, the equilibrium position is stochastically asymptotically stable (resp. stochastically stable, resp. stochastically unstable) if and only if

$$\lim_{t \to \infty} \sup \left(\int_{t_0}^{t} (A(s) - B(s)^2/2) \, ds + \int_{t_0}^{t} B(s) \, dW_s \right)$$

is $-\infty$ almost certainly (resp. is less than ∞ almost certainly, resp. is ∞ with positive probability). Now, in accordance with remark (5.2.6), we can write

$$\int_{t_0}^{t} B(s) \, dW_s = \overline{W}_{\tau(t)}, \quad \tau(t) = \int_{t_0}^{t} B(s)^2 \, ds,$$

where $\overline{W}_t$ is again a Wiener process. In the case

$$\tau(\infty) = \int_{t_0}^{\infty} B(s)^2 \, ds < \infty$$

we therefore have as a criterion

$$\limsup_{t \to \infty} \int_{t_0}^{t} A(s) \, ds = \begin{cases} -\infty \\ < \infty \\ +\infty \end{cases}$$

(the behavior of the undisturbed equation), characterizing stochastic asymptotic stability (resp. stability, resp. instability).

In the case

$$\tau(\infty) = \int_{t_0}^{\infty} B(s)^2 \, ds = \infty$$

let us consider the quantity

$$J(t) = \frac{\displaystyle\int_{t_0}^{t} (A(s) - B(s)^2/2) \, ds}{\sqrt{2 \, \tau(t) \log \log \tau(t)}}.$$

By virtue of the law of the iterated logarithm for $\overline{W}_{\tau(t)}$, we see that a sufficient condition for stochastic asymptotic stability (resp. stochastic instability) is that $\limsup J(t) < -1$ (resp. $\liminf J(t) > -1$).

(11.4.9) **Remark.** For the solution $X_t(c)$ of (11.4.1), we have $X_t(\alpha c) = \alpha X_t(c)$ for all $\alpha \in R^1$. Therefore,

$$P\left[\lim_{t \to \infty} X_t(c) = 0\right] = p_c = \text{const}, \text{for all } c \in R^d.$$

In the case of stochastic asymptotic stability of the equilibrium position, $p_c \to 1$ as $c \to 0$. This is compatible only with $p_c \equiv 1$, hence with

$$\text{ac-}\lim_{t \to \infty} X_t(c) = 0 \text{ for all } c \in R^d,$$

that is, if the equilibrium position of a linear equation is stochastically asymptotically stable, it must automatically be stochastically asymptotically stable in the large.

The extension L (introduced in (11.2.3)) of the infinitesimal operator A of X_t assumes, for the linear equation (11.4.1), the following form:

$$L = \frac{\partial}{\partial t} + \left(A(t) x, \frac{\partial}{\partial x} \right) + \frac{1}{2} \sum_{i=1}^{m} \left(B_i(t) x, \frac{\partial}{\partial x} \right)^2,$$

(11.4.10)
$$\left(y, \frac{\partial}{\partial x} \right) = \sum_{j=1}^{d} y_j \frac{\partial}{\partial x_j},$$

$$\left(y, \frac{\partial}{\partial x} \right)^2 = \sum_{j=1}^{d} \sum_{k=1}^{d} y_j y_k \frac{\partial^2}{\partial x_j \, \partial x_k}.$$

We now cite a criterion for exponential stability in mean square, for the proof of which for general $p > 0$ we refer the reader to Khas'minskiy [65], p. 247.

(11.4.11) **Theorem.** Suppose that the functions $A(t)$ and $B_i(t)$ in equation (11.4.1) are bounded on $[t_0, \infty)$. Then, a necessary (resp. sufficient) condition for exponential stability in mean square is that, for every (resp. some) symmetric positive-definite continuous bounded $d \times d$ matrix $C(t)$ such that $x' C(t) x \geqq k_1 |x|^2$ (where $k_1 > 0$), the matrix differential equation

(11.4.12) $$\frac{dD(t)}{dt} + A(t)' D(t) + D(t) A(t) + \sum_{i=1}^{m} B_i(t)' D(t) B_i(t) = -C(t)$$

have as its solution a matrix $D(t)$ with the same properties as $C(t)$.

(11.4.13) **Remark.** When $B_i(t) \equiv 0$ (for all i), Theorem (11.4.11) reduces to a criterion (well-known in the study of deterministic equations) for ordinary exponential stability (see Bucy and Joseph [61], p. 11). Equation (11.4.12), which has a structural similarity to (11.4.3), can be written with the quadratic forms $w(t, x) = x' C(t) x$ and $v(t, x) = x' D(t) x$ and the operator L in the form

$$L v(t, x) = -w(t, x)$$

that is, $v(t, x)$ can be used as Lyapunov function. Theorem (11.2.8c) again yields stochastic asymptotic stability in the large as a consequence of the exponential stability of the second moments.

(11.4.14) **Corollary.** If equation (11.4.1) is *autonomous*, a necessary (resp. sufficient) condition for exponential stability in mean square is that, for every (resp. for some) symmetric positive-definite matrix C, the matrix equation

(11.4.15) $$A' D + D A + \sum_{i=1}^{m} B_i' D B_i = -C$$

have a symmetric positive-definite solution D. If this is the case, we also have stochastic asymptotic stability in the large.

For verification of this criterion, we begin with any positive-definite matrix C (for example $C = I$), calculate D from the system (11.4.15) of $d(d+1)/2$ linear equations, and check to see whether the D found is positive-definite or not.

(11.4.16) **Example.** If for the autonomous equation

$$dX_t = A X_t \, dt + \sum_{i=1}^{m} B_i X_t \, dW_t^i$$

we have $A + A' = c_1 I$ and $B_i = d_i I$, then $D = I$ is a solution of (11.4.15) for $C = -(c_1 + \sum d_i^2) I$. This last matrix is positive-definite if and only if $c_1 + \sum d_i^2 < 0$.

(11.4.17) **Remark.** Exponential stability in pth mean (for $p > 0$) implies stochastic asymptotic stability of the equilibrium position of the linear equation (11.4.1). Conversely, we have for the *autonomous* linear equation (see Khas'-minskiy [65], pp. 253-257): If the equilibrium position is stochastically asymptotically stable, it is asymptotically stable in pth mean for all sufficiently small $p > 0$. Just as in the deterministic case, this last fact always implies exponential stability in pth mean.

11.5 The Disturbed nth-Order Linear Equation

The equilibrium position of the deterministic nth-order linear equation with constant coefficients

$$y^{(n)} + b_1 y^{(n-1)} + \cdots + b_n y = 0$$

is, according to the Routh-Hurwitz criterion, asymptotically stable if and only if

$$\Delta_1 = b_1 > 0,$$

$$\Delta_2 = \begin{vmatrix} b_1 & b_3 \\ 1 & b_2 \end{vmatrix} > 0,$$

$$\vdots$$

$$\Delta_n = \begin{vmatrix} b_1 & b_3 & b_5 & \cdots & 0 \\ 1 & b_2 & b_4 & \cdots & 0 \\ 0 & b_1 & b_3 & \cdots & 0 \\ \vdots & \vdots & \vdots & & \vdots \\ 0 & \cdots & & \cdots & b_n \end{vmatrix} > 0.$$

The corresponding equation with noisy coefficients

$$Y_t^{(n)} + (b_1 + \xi_1(t)) Y_t^{(n-1)} + \cdots + (b_n + \xi_n(t)) Y_t = 0,$$

where $\xi_1(t), \ldots, \xi_n(t)$ are in general correlated Gaussian white noise processes with

$$E\,\xi_i(t)\,\xi_j(s) = Q_{ij}\,\delta(t-s)$$

is now rewritten, just as in example (8.1.4), as a stochastic first-order differential equation for the n-dimensional process

$$X_t = \begin{pmatrix} X_t^1 \\ \vdots \\ X_t^n \end{pmatrix} = \begin{pmatrix} Y_t \\ \dot{Y}_t \\ \vdots \\ Y_t^{(n-1)} \end{pmatrix}.$$

We obtain

$$dX_t^i = X_t^{i+1}\,dt, \quad i = 1, \ldots, n-1,$$

(11.5.1)

$$dX_t^n = -\sum_{i=1}^{n} b_i\,X_t^{n+1-i}\,dt - \sum_{i=1}^{n}\sum_{j=1}^{n} G_{ij}\,X_t^{n+1-i}\,dW_t^j$$

with an $n \times n$ matrix G such that $G\,G' = Q$.

In accordance with the criteria of section 11.4 (the Routh-Hurwitz criterion for equation (11.4.4) or Corollary (11.4.14)), to prove the asymptotic (= exponential) stability in mean square of (11.5.1), we must treat a system of $n(n+1)/2$ linear equations. We now cite a criterion of Khas'minskiy ([65], pp. 286-292) that operates with only $n+1$ decisions.

(11.5.2) **Theorem.** The equilibrium position of (11.5.1) is asymptotically stable in mean square if and only if $\Delta_1 > 0, \ldots, \Delta_n > 0$ (the Δ_i are the Routh-Hurwitz determinants mentioned above; their positiveness implies stability of the undisturbed equation) and $\Delta_n > \Delta/2$, where

$$\Delta = \begin{vmatrix} q_{nn}^{(0)} & q_{nn}^{(1)} & q_{nn}^{(2)} & \cdots & q_{nn}^{(n-1)} \\ 1 & b_2 & b_4 & \cdots & 0 \\ 0 & b_1 & b_3 & \cdots & 0 \\ \vdots & \vdots & \vdots & & \vdots \\ 0 & 0 & 0 & \cdots & b_n \end{vmatrix},$$

$$q_{nn}^{(n-k-1)} = \sum_{i+j=2(n-k)} Q_{ij}\,(-1)^{j+1}.$$

For $n = 2$, Theorem (11.5.2) provides the conditions

$$b_1 > 0, \quad b_2 > 0, \quad 2\,b_1\,b_2 > Q_{11}\,b_2 + Q_{22},$$

(see example (11.4.5)). For $n = 3$,

$$b_1 > 0, \quad b_3 > 0, \quad b_1 b_2 > b_3,$$

$$2(b_1 b_2 - b_3) b_3 > Q_{11} b_2 b_3 + Q_{33} b_1 + b_3 (a_{22} - 2 Q_{13}).$$

In the case of uncorrelated disturbances ($Q_{ij} = 0$ for $i \neq j$), we have

$$\Delta = \begin{vmatrix} Q_{11} & -Q_{22} & \cdots & (-1)^{n-1} Q_{nn} \\ 1 & b_2 & \cdots & 0 \\ 0 & b_1 & \cdots & 0 \\ \vdots & \vdots & & \vdots \\ 0 & 0 & \cdots & b_n \end{vmatrix}.$$

11.6 Proof of Stability by Linearization

Just as in the deterministic case, stability of a nonlinear equation is in general difficult to prove. However, the proof is facilitated by the fact that the equation linearized by means of a Taylor expansion usually exhibits in the neighborhood of the equilibrium position the same behavior as regards stability as does the original equation. We mention the following theorem of Khas'minskiy ([65], p. 299):

(11.6.1) **Theorem.** Suppose that

$$(11.6.2) \qquad dX_t = f(t, X_t) \, dt + G(t, X_t) \, dW_t, \quad X_{t_0} = c,$$

is a stochastic differential equation for which the assumptions (11.2.1) hold. Let us suppose that

$$|f(t, x) - A(t) x| = o(|x|)$$

and

$$|G(t, x) - (B_1(t) x, \ldots, B_m(t) x)| = o(|x|),$$

uniformly in $t \geq t_0$ as $|x| \to 0$. Let $A(t)$ and $B_i(t)$ denote $d \times d$ matrices that are bounded functions of t. Consider the linear equation

$$(11.6.3) \qquad dX_t = A(t) X_t \, dt + \sum_{i=1}^{m} B_i(t) X_t \, dW_t^i, \quad X_{t_0} = c.$$

a) Suppose that, for the equilibrium position of (11.6.3),

$$\lim_{c \to 0} P \left[\sup_{s \leq t < \infty} |X_t(s, x)| > \varepsilon \right] = 0 \quad \text{for all } \varepsilon > 0,$$

uniformly in s and that $P\left[\lim_{t \to \infty} X_t(s, x) = 0\right] = 1$ (uniform stochastic stability in the large). Then the equilibrium position of (11.6.2) is stochastically asymptotically stable.

b) If the matrices A and B_i are independent of t, then stochastic asymptotic stability of the equilibrium position of the linearized equation (11.6.3) implies stochastic asymptotic stability of the original equation (11.6.2).

11.7 An Example From Satellite Dynamics

The study of the influence of a rapidly fluctuating density of the atmosphere of the earth on the motion of a satellite in a circular orbit leads to the equation (we follow Sagirow [75])

$$(11.7.1a) \quad \ddot{Y}_t + B\,(1 + A\,\xi_t)\,\dot{Y}_t + (1 + A\,\xi_t)\,\sin Y_t - C\,\sin(2\,Y_t) = 0,$$

where B and C are positive numbers and ξ_t is a scalar white noise. The equivalent stochastic differential equation (with $d=2$ and $m=1$) for $X_t = (Y_t,\ \dot{Y}_t)'$ is

$$(11.7.1b) \quad dX_t = \begin{pmatrix} X_t^2 \\ -\sin X_t^1 + C\,\sin 2\,X_t^1 - B\,X_t^2 \end{pmatrix} dt + \begin{pmatrix} 0 \\ -A\,(\sin X_t^1 + B\,X_t^2) \end{pmatrix} dW_t.$$

Replacement of $\sin y$ (resp. $\sin 2\,y$) with y (resp. $2\,y$) yields the linearized equation with constant coefficients

$$(11.7.2a) \quad \ddot{Y}_t + B\,(1 + A\,\xi_t)\,\dot{Y}_t + (1 - 2\,C + A\,\xi_t)\,Y_t = 0$$

in the first case and

$$(11.7.2b) \quad dX_t = \begin{pmatrix} 0 & 1 \\ 2\,C-1 & -B \end{pmatrix} X_t\,dt + \begin{pmatrix} 0 & 0 \\ -A & -A\,B \end{pmatrix} X_t\,dW_t$$

in the second. A necessary and sufficient condition for asymptotic (and hence exponential) stability in mean square of the linearized equation (11.7.2) is, according to the criterion (11.5.2) with

$$Q = \begin{pmatrix} B^2\,A^2 & B\,A^2 \\ B\,A^2 & A^2 \end{pmatrix}$$

and with $d = n = 2$, the following:

$$B > 0, \quad 1 - 2\,C > 0, \quad 2\,B\,(1 - 2\,C) > B^2\,A^2\,(1 - 2\,C) + A^2.$$

The conditions $B > 0$ and $1 - 2\,C > 0$ ensure asymptotic stability of the undisturbed system as a necessary condition. The last condition yields the inequality

$$A^2 < \frac{2\,B\,(1 - 2\,C)}{B^2\,(1 - 2\,C) + 1}.$$

for the intensity of the disturbance. Therefore, under these conditions, the equilibrium position of (11.7.2)—hence, by Theorem (11.6.1), the equilibrium position of the nonlinear equation—is stochastically asymptotically stable.

We now seek to use Lyapunov's technique to carry these assertions over to the nonlinear original equation. The operator L assumes for (11.7.1) the following form:

$$L = \frac{\partial}{\partial t} + x_2 \frac{\partial}{\partial x_1} + (-\sin x_1 + C \sin 2 x_1 - B x_2) \frac{\partial}{\partial x_2}$$
$$+ \frac{A^2}{2} (\sin x_1 + B x_2)^2 \frac{\partial^2}{\partial x_2^2}.$$

For the Lyapunov function, we try an expression consisting of a quadratic form and integrals of the nonlinear components:

$$v(t, x) \equiv v(x) = a x_1^2 + b x_1 x_2 + x_1^2 + d \int_0^{x_1} \sin y \, dy + e \int_0^{x_1} \sin 2 y \, dy$$

$$= a x_1^2 + b x_1 x_2 + x_2^2 + 2 d \left(\sin \frac{x_1}{2} \right)^2 + e (\sin x_1)^2.$$

This yields

$$L v(x) = (2 a - b B) x_1 x_2 - (2 B - b - A^2 B^2) x_2^2$$
$$+ (d - 2 + 2 A^2 B + (4 C + 2 e) \cos x_1) x_2 \sin x_1$$
$$- \left(b - 2 b C \cos x_1 - A^2 \frac{\sin x_1}{x_1} \right) x_1 \sin x_1.$$

To convert this to a negative-definite function, we set $2 a - b B = 0$, $4 C + 2 e = 0$, and $d - 2 + 2 A^2 B = 0$. We obtain

$$v(x) = \frac{1}{2} b B x_1^2 + b x_1 x_2 + x_2^2 + 4 \left(\sin \frac{x_1}{2} \right)^2 \left(1 - B A^2 - 2 C \left(\cos \frac{x_1}{2} \right)^2 \right)$$

and

$$L v(x) = -(2 B - b - A^2 B^2) x_2^2 - \left(b - 2 b C \cos x_1 - A^2 \frac{\sin x_1}{x_1} \right) x_1 \sin x_1.$$

For v to be positive-definite, we must have $B > 0$, $0 < b < 2 B$, $1 - 2 C > 0$, and $A^2 < (1 - 2 C)/B$. Furthermore, we can get the following estimate for $L v$:

$$L v(x) \leqq -(2 B - b - A^2 B^2) x_2^2 - (b - 2 b C - A^2) x_1 \sin x_1.$$

This last is negative-definite if $A^2 < (2 B - b)/B^2$ and $A^2 < b (1 - 2 C)$ and hence if

$$A^2 < \min ((2 B - b)/B^2, \, b (1 - 2 C)).$$

If we choose b in the interval $0 < b < 2 B$ in such a way that the minimum is as large as possible, we get

$$b = \frac{2\,B}{1 + B^2\,(1 - 2\,C)}$$

and hence, for the intensity of the disturbance,

$$A^2 < \frac{2\,B\,(1 - 2\,C)}{B^2\,(1 - 2\,C) + 1}.$$

Theorem (11.2.8b) ensures, under this condition, the stochastic asymptotic stability of the equilibrium position of (11.7.1). This result is identical to the result obtained by linearization.

Chapter 12

Optimal Filtering of a Disturbed Signal

12.1 Description of the Problem

If the state X_t of a stochastic dynamic system is used to make a decision (see Chapter 13), we naturally assume that this state is exactly known to us. Such an assumption, however, is unrealistic in many practical problems. What can be observed on the basis of technical or economic considerations is a process Z_t that in some way depends on the previous behavior of X_t and, in addition, is disturbed.

The question then arises as to how one can, in an optimal way (i.e. with the smallest possible error), draw a conclusion on the basis of observation of Z_t as to the true state X_t of the system in question. For the case of stationary linear systems, this problem was solved in the 1940's independently by N. Wiener [79] and A. N. Kolmogorov [69] (see also, for example, Gikhman and Skorokhod [5] or Prohorov and Rozanov [15]).

Nonlinear systems and the case of observations made only during a finite interval have been treated since 1960 by R. E. Kalman, R. S. Bucy, R. L. Stratonovich, and H. J. Kushner. We shall now discuss their results, confining ourselves to the basic ideas. For additional results and references in the literature, we mention Bucy and Joseph [61] and the detailed book by Jazwinski [66].

The basis of our studies is the following class of models:

Suppose that the d-dimensional state X_t of a dynamic system (the so-called **signal process**) is described by a stochastic differential equation

$$(12.1.1) \quad dX_t = f(t, X_t)\, dt + G(t, X_t)\, dW_t, \quad X_{t_0} = c \in L^2, \quad t \geqq t_0.$$

Here, $f(t, x)$ is again a d-dimensional vector, $G(t, x)$ is a $d \times m$ matrix, and W_t is an m-dimensional Wiener process whose "derivative" is therefore a white noise, that is, a generalized stationary Gaussian process with mean value 0 and covariance matrix $I\, \delta(t - s)$. As usual, we reduce the case (often encountered in the literature) in which the derivative of W_t is a stationary process with covariance matrix $Q(t)\, \delta(t - s)$ to (12.1.1) by shifting from G to $G\sqrt{Q}$.

Suppose that the *observed process* Z_t (the quantity being measured) is p-dimensional and that it is a disturbed functional of X_t of the form

(12.1.2) $dZ_t = h(t, X_t)\, dt + R(t)\, dV_t, \quad Z_{t_0} = b, \quad t \geq t_0.$

Here, $h(t, x)$ is a p-vector, $R(t)$ is a $p \times q$ matrix, and V_t is a q-dimensional Wiener process. We assume that the four random objects $W_.$, $V_.$, c, and b are independent. For correlated noise processes W_t and V_t, one should compare, for example, Kalman [67] and Kushner [71]. Of course, X_t and Z_t are dependent in any case.

We also assume that equation (12.1.1) satisfies the assumptions of the existence-and-uniqueness theorem (6.2.1), so that there exists a global solution in the interval $[t_0, \infty)$. Equation (12.1.2) should be understood as meaning that, when we substitute the solution X_t of (12.1.1) into the argument of h, this renders the right-hand member of (12.1.2) an ordinary stochastic differential in the sense of section 5.3 since Z_t does not appear in that member. To be sure that (12.1.2) is always meaningful, we must require that

$$\int_{t_0}^{t} |R(s)|^2\, ds < \infty \text{ for all } t > t_0,$$

and, for example,

$$|h(t, x)| \leq C(1 + |x|^r) \text{ for all } t \geq t_0, \quad x \in R^d,$$

for some $r > 0$.

Now, the filtering problem can be formulated as follows:

(12.1.3) **Problem.** Consider the observed values Z_s, where $t_0 \leq s \leq t$. For this piece of a function we write $Z[t_0, t]$. Suppose that $t_1 > t_0$. Let us construct a d-dimensional random variable $\widehat{X}_{t_1}$ as a measurable functional of $Z[t_0, t]$ such that, for every other measurable functional $F(Z[t_0, t])$ with range in R^d such that

$$E\,\widehat{X}_{t_1} = E\,F(Z[t_0, t]) = E\,X_{t_1} \text{ (unbiasedness)}$$

the inequality

(12.1.4) $E\,(y'(X_{t_1} - \widehat{X}_{t_1}))^2 \leq E\,(y'(X_{t_1} - F(Z[t_0, t])))^2 \text{ for all } y \in R^d,$

holds. The quantity $\widehat{X}_{t_1}$ is called the **(optimal) estimate** of X_t for the given observation $Z[t_0, t]$. For $t_1 < t$, we speak of **interpolation**, for $t_1 = t$, we speak of **smoothing** or **filtering**; and for $t_1 > t$, we speak of **extrapolation** or **prediction**. By virtue of the optimality criterion (12.1.4), we speak of the **method of minimum variance**.

(12.1.5) **Remark.** It should be noted that the value $\widehat{X}_{t_1}(\omega)$ may depend on the total trajectory $Z_.(\omega)$ of the process observed in the time interval $[t_0, t]$ and it

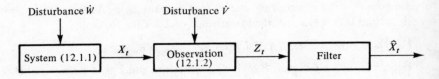

Fig. 8: The scheme of optimal filtering.

must, conversely, be uniquely determined by this piece of trajectory. Measurability of the functional $F\,(Z\,[t_0,\,t])$ has the following meaning: by virtue of remark (5.3.2), the sample functions $Z_.\,(\omega)$ of the observed process are, with probability 1, continuous functions. If we neglect the ω-exceptional set of probability 0 (or if we take, for example $Z_t\,(\omega)\equiv 0$ on that set), then the mapping $Z\,[t_0,\,t]$ that assigns to $\omega\in\Omega$ the corresponding piece of the trajectory $Z_s\,(\omega)$ for $t_0\leqq s\leqq t$ is a mapping of Ω into the space $C\,([t_0,\,t])$ of continuous functions on the interval $[t_0,\,t]$:

(12.1.6) $Z\,[t_0,\,t]:\Omega\;\longrightarrow\;C\,([t_0,\,t]).$

The sigma-algebra $\mathfrak{A}$ is defined in Ω. If we choose in $C\,([t_0,\,t])$ the sigma algebra $\mathfrak{C}\,([t_0,\,t])$ generated by the "spheres"

$$\{\varphi\in C\,([t_0,\,t]):\max_{t_0\leqq s\leqq t}\,|\varphi\,(s)-\psi\,(s)|<\varepsilon\},\quad \psi\in C\,([t_0,\,t]),\quad \varepsilon>0,$$

(ε-neighborhoods of the fixed continuous function ψ), then the mapping (12.1.6) is $\mathfrak{A}-\mathfrak{C}\,([t_0,\,t])$-measurable. A functional that is measurable in the sense of the above-given formulation of the problem is a mapping

$$F:\;C\,([t_0,\,t])\;\longrightarrow\;R^d$$

that is measurable with respect to $\mathfrak{C}\,([t_0,\,t])$ and the Borel sigma-algebra $\mathfrak{B}^d$ in R^d. Then, of course, the combined mapping

$$F\,(Z\,[t_0,\,t]):\Omega\;\longrightarrow\;R^d$$

is a d-dimensional random variable that depends on ω only through $Z\,[t_0,\,t]$. We now seek a measurable $F_0:\,C\,([t_0,\,t])\rightarrow R^d$ such that

$$\widehat{X}_{t_1}=F_0\,(Z\,[t_0,\,t])$$

has a second moment and expectation value $E\,X_{t_1}$ and possesses the minimality property (12.1.4).

(12.1.7) **Remark.** The condition (12.1.4) is equivalent to the following condition: For every nonnegative-definite symmetric matrix C, we have the inequality

$$E\,|X_{t_1}-\widehat{X}_{t_1}|_C^2\leqq E\,|X_{t_1}-F\,(Z\,[t_0,\,t])|_C^2.$$

with the abbreviation $x'\,C\,x=|x|_C^2$ for $x\in R^d$. The equivalence with (12.1.4) is seen from the spectral decomposition of C:

$$C = \sum_{i=1}^{d} \lambda_i (u_i \, u_i'); \; \lambda_i \text{ and } u_i \text{ are the eigenvalues and eigenvectors of } C,$$

the relationship

$$|x|_C^2 = \sum_{i=1}^{d} \lambda_i (u_i' \, x)^2$$

and the choice of the special matrix $C = y \, y'$. We mention that the requirement

$$E \, |X_{t_1} - \widehat{X}_{t_1}|^2 \leqq E \, |X_{t_1} - F \, (Z \, [t_0, t])|^2$$

is for $d > 1$ weaker than (12.1.4). This last also includes the nondiagonal members of the matrix $E \, (X_{t_1} - \widehat{X}_{t_1}) \, (X_{t_1} - \widehat{X}_{t_1})'$ in the comparison.

12.2 The Conditional Expectation as Optimal Estimate

The existence and uniqueness of the optimal estimate are relatively easy to prove.

(12.2.1) **Theorem.** The unique solution (up to a set of probability 0) of the problem (12.1.3) is

$$\widehat{X}_{t_1} = E \, (X_{t_1} | Z \, [t_0, t]).$$

Therefore (see section (1.7)), there exists a measurable function F_0 defined on $C \, ([t_0, t])$, unique on the set of possible trajectories $Z \, [t_0, t]$, such that

$$\widehat{X}_{t_1} = F_0 \, (Z \, [t_0, t]).$$

Proof. For an arbitrary estimate $\widehat{X}_{t_1}$,

$$E \, (y' \, (X_{t_1} - F \, (Z \, [t_0, t])))^2 = E \, (y' \, (X_{t_1} - \widehat{X}_{t_1}))^2 + E \, (y' \, (\widehat{X}_{t_1} - F))^2$$
$$+ 2 \, y' \, E \, (X_{t_1} - \widehat{X}_{t_1}) \, (\widehat{X}_{t_1} - F)' \, y.$$

On the other hand, in accordance with section 1.7, we have

$$E \, (X_{t_1} - \widehat{X}_{t_1}) \, (\widehat{X}_{t_1} - F)' = E \, (E \, ((X_{t_1} - \widehat{X}_{t_1}) \, (\widehat{X}_{t_1} - F)' | Z \, [t_0, t]))$$
$$= E \, (E \, ((X_{t_1} - \widehat{X}_{t_1}) | Z \, [t_0, t]) \, (\widehat{X}_{t_1} - F)')$$
$$= 0,$$

if we set

$$\widehat{X}_{t_1} = E \, (X_{t_1} | Z \, [t_0, t]) \, .$$

Then, for this $\widehat{X}_{t_1}$,

$$E \, (y' \, (X_{t_1} - \widehat{X}_{t_1}))^2 \leqq E \, (y' \, (X_{t_1} - F))^2,$$

so that the conditional expectation is a solution of the problem. The uniqueness follows from the geometrical fact that $E(X_{t_1}|Z[t_0, t])$ in the Hilbert space $L^2(\Omega, \mathfrak{A}, P)$ is the unique orthogonal projection of X_{t_1} onto the linear subspace of those elements that are measurable with respect to the sigma-algebra generated by $Z[t_0, t]$ (see Krickeberg [7], pp. 124-125). ∎

For the value of the functional F_0 in Theorem (12.2.1) at the "point" $\varphi \in C([t_0, t])$, we write suggestively

$$F_0(\varphi) = E(X_{t_1}[Z[t_0, t] = \varphi), \quad \varphi \in C([t_0, t]).$$

In particular, if $t = t_0$, we have, by virtue of the independence of $Z_{t_0} = b$ and X,

$$\widehat{X}_{t_1} = E(X_{t_1}|Z_{t_0}) = E X_{t_1} = \text{const.}$$

From here on, we shall concern ourselves only with the **filtering problem** $(t_1 = t)$. Theoretically, it is completely solved by (12.2.1), a fact that, however, does not satisfy us. What we are seeking is an algorithm with the aid of which we can calculate the numerical value of the optimal estimate $\widehat{X}_t$ of the state X_t of the system from the numerical observations Z_s for $t_0 \leq s \leq t$. (By virtue of the continuity of Z., it will be sufficient to make observations in (t_0, t).) This calculation may, apart from the observation itself, use only the system parameters f, G, h, and R and the initial distribution P_{t_0} of $X_{t_0} = c$.

12.3 The Kalman-Bucy Filter

We assume the same situation as in section 12:1. Throughout, $\widehat{X}_t$ will denote the conditional expectation. If X_t has a conditional density $p_t(x|Z[t_0, t])$ under the condition $Z[t_0, t]$, then, in accordance with section 1.7,

$$\widehat{X}_t = E(X_t|Z[t_0, t]) = \int_{R^d} x\, p_t(x|Z[t_0, t])\, dx;$$

or, more generally, for $g(x)$ with $g(X_t) \in L^1$,

$$\widehat{g(X_t)} = E(g(X_t)|Z[t_0, t]) = \int_{R^d} g(x)\, p_t(x|Z[t_0, t])\, dx.$$

Therefore, if we know the conditional density $p_t(x|Z[t_0, t])$, we can easily find the optimal estimate $\widehat{X}_t$ by means of an ordinary integration.

(12.3.1) **Theorem (Bucy's representation theorem).** Suppose that we are given equations (12.1.1) and (12.1.2) with the assumptions stated in section 12.1. We also assume that $R(t) R(t)'$ is positive-definite, that the finite-dimensional distributions of the process X_t have densities, and that

$$E \exp\left((t - t_0) \sup_{t_0 \leq s \leq t} |h(s, X_s)|^2 (R(s) R(s)')^{-1}\right) < \infty .$$

Here, we have set $|x|_A^2 = x' A x$. Then, the conditional distribution $P(X_t \in B | Z [t_0, t])$ has a density $p_t (x | Z [t_0, t])$ that, for a fixed observation $Z [t_0, t]$, is given by

$$(12.3.2) \qquad p_t (x | Z [t_0, t]) = \frac{E (e^Q | X_t = x) p_t (x)}{E \, e^Q}.$$

Here, $p_t (x)$ is the density of X_t and

$$(12.3.3) \qquad \begin{aligned} Q &= Q (X [t_0, t], Z [t_0, t]) \\ &= -\frac{1}{2} \int_{t_0}^{t} h (s, X_s)' (R (s) R (s)')^{-1} h (s, X_s) \, ds + \\ &\quad + \int_{t_0}^{t} h (s, X_s)' (R (s) R (s)')^{-1} \, dZ_s. \end{aligned}$$

Formula (12.3.2) was conjectured by Bucy [60] and proven by Mortensen [74] (see also Bucy and Joseph [61]). For *fixed* $Z [t_0, t]$, the integral with respect to Z_s in (12.3.3) is, in complete analogy with the integral with respect to W_t, defined by the approximation of the integrand by means of step functions and the choice of the left end-points of a decomposition as the intermediate points.

For useless observations ($h \equiv 0$ or $(R (t) R (t)')^{-1} \equiv 0$), we obtain from Theorem (12.3.1) $Q \equiv 0$ and hence

$$p_t (x | Z [t_0, t]) = p_t (x),$$

so that

$$\hat{X}_t = E X_t.$$

Similarly, for $t = t_0$, we always have

$$p_{t_0} (x | Z_{t_0}) = p_{t_0} (x) \quad \text{(density of the initial value } c\text{)},$$

that is,

$$\hat{X}_{t_0} = E c.$$

By applying Itô's theorem to the representation (12.3.2), we obtain (see Bucy and Joseph [61]) the so-called **fundamental equation of filtering theory**:

$$(12.3.4) \qquad \begin{aligned} d p_t (x | Z [t_0, t]) &= \mathfrak{D}^* (p_t (x | Z [t_0, t])) \, dt \\ &\quad + (h (t, x) - \hat{h} (t, x))' (R (t) R (t)')^{-1} (dZ_t - \hat{h} (t, x) \, dt) p_t (x | Z [t_0, t]). \end{aligned}$$

Here, $\mathfrak{D}^*$ is the adjoint operator (forward operator) to the differential operator $\mathfrak{D}$:

$$\mathfrak{D}^* g = -\sum_{i=1}^{d} \frac{\partial}{\partial x_i} (f_i \, g) + \frac{1}{2} \sum_{i=1}^{d} \sum_{j=1}^{d} \frac{\partial^2}{\partial x_i \, \partial x_j} ((G G')_{ij} \, g) ,$$

and

$$\widehat{h}\,(t,\,x) = E\,(h\,(t,\,X_t)|Z\,[t_0,\,t]) = \int_{R^d} h\,(t,\,x)\,p_t\,(x|Z\,[t_0,\,t])\,\mathrm{d}x.$$

Equation (12.3.4) shows the possibility, at least in theory, of calculating the conditional density $p_t\,(x|Z\,[t_0,\,t])$, beginning with the initial value $p_{t_0}\,(x)$ with progressive observation of $Z_{\cdot}$. Here, $p_t\,(x|Z\,[t_0,\,t])$ depends only on the two systems defined by the functions f, G, h, and R, on the density of the initial value, and on the observations up to the instant t. Therefore, (12.3.4) can be regarded as the dynamic equation for the optimal filter.

Equation (12.3.4) yields, upon integration with respect to x, equations for the moments of the conditional density, in particular, for the optimal estimate $\widehat{X}_t$:

$$\mathrm{d}\widehat{X}_t = \widehat{f}\,(t,\,x)\,\mathrm{d}t + (\widehat{x\,h}\,(t,\,x)' - \widehat{X}_t\,\widehat{h}\,(t,\,x)')\,(R\,(t)\,R\,(t)')^{-1}$$
$$(\mathrm{d}Z_t - \widehat{h}\,(t,\,x)\,\mathrm{d}t)$$

with the initial value $\widehat{X}_{t_0} = E\,c$. Also of interest is the **estimation error**, that is, the conditional covariance matrix

$$P\,(t|Z\,[t_0,\,t]) = E\,((X_t - \widehat{X}_t)\,(X_t - \widehat{X}_t)'|Z\,[t_0,\,t]).$$

For this, we get from (12.3.4) (see Jazwinski [66], p. 184)

$$\mathrm{d}\,(P\,(t|Z\,[t_0,\,t]))_{ij} = ((\widehat{x_i\,f_j}) - \widehat{x}_i\,\widehat{f}_j) + (\widehat{f_i\,x_j} - \widehat{f}_i\,\widehat{x}_j) + (\widehat{G\,G'})_{ij}$$
$$- (\widehat{x_i\,h} - \widehat{x}_i\,\widehat{h})'\,(R\,R')^{-1}\,(\widehat{h\,x_j} - \widehat{h}\,\widehat{x}_j)\,\mathrm{d}t$$
$$+ (\widehat{x_i\,x_j\,h} - \widehat{x_i\,x_j}\,\widehat{h} - \widehat{x}_i\,\widehat{x_j\,h} - \widehat{x}_j\,\widehat{x_i\,h} + 2\,\widehat{x}_i\,\widehat{x}_j\,\widehat{h})'\,(R\,R')^{-1}\,(\mathrm{d}Z_t - \widehat{h}\,\mathrm{d}t)$$

with the initial value $P\,(t_0|Z_{t_0}) = E\,c\,c'$.

In the following section, we shall specialize these equations to linear problems.

12.4   Optimal Filters for Linear Systems

Suppose that a stochastic differential equation for a d-dimensional signal process X_t is linear (in the narrow sense), that is, of the form

$$\mathrm{d}X_t = A\,(t)\,X_t\,\mathrm{d}t + B\,(t)\,\mathrm{d}W_t, \quad X_{t_0} = c, \quad t \geqq t_0,$$

where $A\,(t)$ is a $d \times d$ matrix, $B\,(t)$ is a $d \times m$ matrix, and W_t is an m-dimensional Wiener process. Suppose that an observed p-dimensional process Z_t is described by an equation that is linear in X_t:

$$\mathrm{d}Z_t = H\,(t)\,X_t\,\mathrm{d}t + R\,(t)\,\mathrm{d}V_t, \quad Z_{t_0} = b, \quad t \geqq t_0,$$

where $H\,(t)$ is a $p \times d$ matrix, $R\,(t)$ is a $p \times q$ matrix such that $R\,(t)\,R\,(t)'$ is positive-definite, and V_t is a q-dimensional Wiener process. Suppose that the matrix functions $A(t)$, $B(t)$, $H(t)$, $R(t)$, and $(R(t) R(t)')^{-1}$ are bounded in every bounded subinterval of $[t_0, \infty)$. We again assume the independence of $W_.$, $V_.$, c, and b. If c and b are normally distributed or constant, then, in accordance with Theorem (8.2.10), X_t and hence Z_t are Gaussian processes. Therefore, all the conditional distributions are normal distributions. In particular, we have

(12.4.1) Theorem (Kalman-Bucy filters for linear systems). In the linear case, the conditional density $p_t(x|Z[t_0, t])$ of X_t under the condition that $Z[t_0, t]$ was observed is the density of a normal distribution with mean

$$\hat{X}_t = E(X_t|Z[t_0, t])$$

and covariance matrix

$$P(t) = E((X_t - \hat{X}_t)(X_t - \hat{X}_t)'|Z[t_0, t]) = E(X_t - \hat{X}_t)(X_t - \hat{X}_t)'.$$

The dynamic equations for these parameters are

$$d\hat{X}_t = A\,\hat{X}_t\,dt + P\,H'\,(R\,R')^{-1}\,(dZ_t - H\,\hat{X}_t\,dt), \quad \hat{X}_{t_0} = E\,c,$$

$$\dot{P} = A\,P + P\,A' + B\,B' - P\,H'\,(R\,R')^{-1}\,H\,P, \quad P(t_0) = E\,c\,c'.$$

In particular, the matrix $P = P(t)$ is independent of the observation $Z[t_0, t]$.

For various proofs of this theorem, see Jazwinski ([66], pp. 218ff).

In the case of a useless or missing observation ($H(t) \equiv 0$ or $(R(t) R(t)')^{-1} \equiv 0$), the equations for $\hat{X}_t$ and $P(t)$ reduce to the equations for the mean $\hat{X}_t = E\,X_t$ and the covariance matrix $P(t) = K(t)$ of the process X_t (Theorem (8.2.6)). Since the coefficient of dZ_t in the equation for $\hat{X}_t$ is deterministic, the corresponding integral can be evaluated without any precautions (arbitrary intermediate points!).

The equation for $P(t)$ is a (matrix) **Riccati equation**. It has been carefully investigated by Bucy and Joseph [61]. In particular, in spite of the quadratic term, there exists a global solution when one starts with a nonnegative-definite initial value $P(t_0)$. Since $P(t)$ is independent of the observation, the estimation error can be evaluated in advance.

(12.4.2) Remark. For a linear model, the optimal estimate $\hat{X}_t$ is identical to the optimal *linear* estimate

$$\tilde{X}_t = \int_{t_0}^{t} D(t, s)\,dZ_s.$$

The $d \times p$ weight matrix $D(t, s)$ is obtained from the so-called **Wiener-Hopf equation**

$$D(t, u) R(u) R(u)' + \int_{t_0}^{t} D(t, s) H(s) E(X_s X_u') H(u)' \, ds$$

$$= E(X_t X_u') H(u)'$$

(see Bucy and Joseph [61], p. 53, or Gikhman and Skorokhod [5], p. 229).

(12.4.3) **Example.** Suppose that the signal process is undisturbed $(B(t) \equiv 0)$ and that it starts with a random $\mathfrak{N}(0, E c c')$-distributed initial value with positive-definite $E c c'$. The (unconditional) distribution of X_t (*before* observation!) is, by Theorem (8.2.10),

$$\mathfrak{N}(0, \Phi(t) E c c' \Phi(t)'),$$

where $\Phi(t)$ is the fundamental matrix of $\dot{X}_t = A(t) X_t$. The conditional distribution of X_t *after* observation of $Z[t_0, t]$ is, as one can verify by substitution into the equations of Theorem (12.4.1), a normal distribution $\mathfrak{N}(\hat{X}_t, P(t))$ with

$$\hat{X}_t = P(t) (\Phi(t)')^{-1} \int_{t_0}^{t} \Phi(s)' H(s)' (R(s) R(s)')^{-1} \, dZ_s$$

(so that $\hat{X}_t$ is a linear estimate in the sense of remark (12.4.2)) and

$$P(t) = \left((\Phi(t) E c c' \Phi(t)')^{-1} \right.$$

$$\left. + (\Phi(t)')^{-1} \left(\int_{t_0}^{t} \Phi(s)' H(s)' (R(s) R(s)')^{-1} H(s) \Phi(s) \, ds \right) \Phi(t)^{-1} \right)^{-1}.$$

Since the second summand in $P(t)$ is positive-definite, the error covariance matrix $P(t)$ is "smaller" than the original covariance matrix of X_t.

Chapter 13
Optimal Control of Stochastic Dynamic Systems

13.1 Bellman's Equation

The analytical difficulties that arise with a mathematically rigorous treatment of stochastic control problems are so numerous that, in this brief survey, we must confine ourselves on didactic grounds to a more qualitative and intuitive treatment.

As in the case of stability theory, there is again here a well-developed theory for deterministic systems, which one can study, for example, in the works by Athans and Falb [56], Strauss [78], or Kalman, Falb, and Arbib [68]. It is again a matter of replacing, in the shift to the stochastic case, the first-order derivatives at the corresponding places with the infinitesimal operator of the corresponding process. For a more detailed treatment of optimal control of stochastic systems, we refer to the books and survey articles by Aoki [55], Kushner [72], Stratonovich [77], Bucy and Joseph [61], Mandl [28], Khas'minskiy [65], Fleming [62], and Wonham [80] and to the literature cited in those works.

Let us now consider a system described by the stochastic differential equation

$$dX_t = f(t, X_t, u(t, X_t))\, dt + G(t, X_t, u(t, X_t))\, dW_t,$$

(13.1.1) $$X_{t_0} = c, \quad t \geq t_0,$$

where, as usual, X_t, $f(t, x, u)$, and c assume values in R^d, $G(t, x, u)$ is $(d \times m)$-matrix-valued, and W_t is an m-dimensional Wiener process. The new variable u in the arguments of f and G varies in some R^p, and the functions $f(t, x, \cdot)$ and $G(t, x, \cdot)$ are assumed to be sufficiently smooth. The function $u(t, x)$ in equation (13.1.1) is a **control function** belonging to a **set $\mathfrak{U}$ of admissible control functions.** We shall confine ourselves here to so-called Markov control functions which depend only on t and on the state X_t at the instant t (and not, for example, on the values of X_s for $s < t$). The system (13.1.1) is also called a "plant".

If we substitute a fixed control function $u \in \mathfrak{U}$ in (13.1.1), we get a stochastic differential equation of the usual form. Now, the set $\mathfrak{U}$ must be narrowed down by boundedness and analytical conditions, which we shall not further specify, in such a way that existence and uniqueness of a solution $X_t = X_t^u$, which now depends on $u \in \mathfrak{U}$, are ensured for the differential equation. The solution that starts at x at the instant s will be denoted by $X_t(s, x) = X_t^u(s, x)$. We shall also write $E\, g\, (X_t(s, x)) = E_{s, x}\, g\, (X_t)$.

Suppose that the *costs* arising from the choice of control function u up to an instant $T < \infty$ are, in the case of a start at x at the instant s,

$$(13.1.2) \quad V^u(s, x) = E_{s, x}\left(\int_s^T k\, (r, X_r, u\, (r, X_r))\, dr + M\, (T, X_T)\right).$$

Here, we shall confine ourselves to fixed-time control. In general, T in (13.1.2) is replaced with a random instant τ, at which the process reaches a specified target set. The functions k and M have respectively nonnegative and real values. The integral in (13.1.2) represents the running costs, and the second term in the large parentheses represents the unique cost for a stop at X_T at the instant T.

We now seek the **optimal control function**, that is, the control function $u^* \in \mathfrak{U}$ that minimizes the costs:

$$V\, (s, x) = V^{u^*}(s, x) = \min_{u \in \mathfrak{U}} V^u\, (s, x).$$

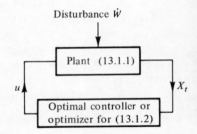

Fig. 9:
Scheme of the optimal control.

In accordance with **Bellman's optimality principle** (the **principle of dynamic programming**, Bellman [57]), a control function defined on the interval $[t_0, T]$ is optimal if and only if it is optimal on every subinterval of the form $[s, T]$, where $t_0 \leqq s < T$. Here, $x = X_s\, (t_0, c)$ is chosen as the initial value at the instant s. In complete analogy with the deterministic case, we then get the result that the minimal costs $V\, (s, x)$ satisfy Bellman's equation

$$(13.1.3) \quad 0 = \min_u (L^u V\, (s, x) + k\, (s, x, u)), \quad t_0 \leqq s \leqq T,$$

with the end condition $V\, (T, x) = M\, (T, x)$. Here,

$$L^u = \frac{\partial}{\partial s} + \sum_{i=1}^d f_i\, (s, x, u)\frac{\partial}{\partial x_i} + \frac{1}{2} \sum_{i=1}^d \sum_{j=1}^d (G\, (s, x, u)\, G\, (s, x, u)')_{ij}\frac{\partial^2}{\partial x_i\, \partial x_j}.$$

The u in the expression L^u is treated as a parameter. In equation (13.1.3),

$L^u V (s, x) + k (s, x, u)$ is, for given V and fixed (s, x), a function of $u \in R^p$, whose minimum is sought. The position u^* of this minimum depends on (s, x); thus, $u^* = u^* (s, x)$. If $V (s, x)$ is equal to the optimal costs and if the function $u^* (s, x)$ resulting from the search for the minimum is an admissible control function, it is also an optimal control function. Then,

$$L^{u^* (s, x)} V (s, x) + k (s, x, u^* (s, x)) = \min_u (L^u V (s, x) + k (s, x, u)) = 0.$$

The following steps therefore yield (under certain conditions) both the optimal control function and the minimum costs:

1. For fixed $\overline{V}$, we determine the point $\bar{u} = \bar{u} (s, x; \overline{V})$ at which $L^u \overline{V} (s, x) + k (s, x, u)$ attains its minimum.

2. We substitute the function $\bar{u}$ for the parameter u in $L^u \overline{V} (s, x) + k (s, x, u)$ and solve the partial differential equation

    $$L^{\bar{u}} V (s, x) + k (s, x, \bar{u} (s, x; V)) = 0, \quad t_0 \leqq s \leqq T,$$

    with the end condition $V (T, x) = M (T, x)$. The solution $V (s, x)$ yields the minimum costs.

3. The function $V (s, x)$ is inserted into the function $\bar{u}$ determined in step 1. It yields the optimal control function $u^* = u^* (s, x) = u (s, x; V (s, x))$.

We shall illustrate this in the next section for the linear case and a quadratic "criterion" (13.1.2).

13.2 Linear Systems

In equation (13.1.1), suppose that $f (t, x, u)$ is linear in x and u and that $G (t, x, u) \equiv G (t)$ depends only on t. We get

$$dX_t = A (t) X_t \, dt + B (t) u (t, X_t) \, dt + G (t) \, dW_t, \quad t \geqq t_0,$$

where $A (t)$ is a $d \times d$ matrix, $B (t)$ is a $d \times p$ matrix, and $G (t)$ is a $d \times m$ matrix. For the functions appearing in the cost functional (13.1.2), we choose

$$k (t, x, u) = x' C (t) x + u' D (t) u,$$

(where $C (t)$ is symmetric and nonnegative-definite and $D (t)$ is symmetric and positive-definite) and

$$M (T, x) = x' F (T) x + a (T)' x + b (T).$$

We have

$$L^u = \frac{\partial}{\partial s} + \left(A (t) x + B (t) u, \frac{\partial}{\partial x} \right) + \frac{1}{2} \sum_{i=1}^{d} \sum_{j=1}^{d} (G (t) G (t)')_{ij} \frac{\partial^2}{\partial x_i \, \partial x_j},$$

so that

$$L^u V(s, x) = \frac{\partial V}{\partial s} + (A(t) x)' V_x + (B(t) u)' V_x + \frac{1}{2} \operatorname{tr}(G(t) G(t)' V_{xx}),$$

where $V_x = (V_{x_1}, \dots, V_{x_d})'$ and $V_{xx} = (V_{x_i x_j})$.

First, we determine the minimizing $\bar{u}$ from Bellman's equation

(13.2.1)
$$\frac{\partial V}{\partial s} + (A(s) x)' V_x + (B(s) u)' V_x + \frac{1}{2} \operatorname{tr}(G(s) G(s)' V_{xx})$$
$$+ x' C(s) x + u' D(s) u = \min.$$

The quadratic function $(B(s) u)' V_x + u' D(s) u$ assumes its minimum when

(13.2.2) $$\bar{u}(s, x; V) = -\frac{1}{2} D(s)^{-1} B(s)' V_x.$$

When we substitute this into (13.2.1), we get the partial differential equation for the minimum costs $V(s, x)$, for $t_0 \leqq s \leqq T$:

(13.2.3) $$\frac{\partial V}{\partial s} + \frac{1}{2} \operatorname{tr}(G G' V_{xx}) + V_x' A x - \frac{1}{4} V_x' B D^{-1} B' V_x + x' C x = 0$$

with the end condition $V(T, x) = x' F(T) x + a(T)' x + b(T)$.

To solve (13.2.3), let us try

$$V(s, x) = x' Q(s) x + q(s)' x + p(s),$$

where $Q(s)$ is symmetric and nonnegative-definite. When we substitute this into (13.2.3) and equate the coefficients, we get, for $t_0 \leqq s \leqq T$, the ordinary (coupled) differential equations for $Q(s)$, $q(s)$, and $p(s)$:

(13.2.4)
$$\dot{Q}(s) + A' Q + Q A + C - Q B D^{-1} B' Q = 0, \quad Q(T) = F(T),$$
$$\dot{q}(s) + (A' - Q B D^{-1} B') q = 0, \quad q(T) = a(T),$$
$$\dot{p}(s) + \operatorname{tr}(G G' Q) - \frac{1}{4} q' B D^{-1} B' q = 0, \quad p(T) = b(T).$$

These must be solved backwards in direction t_0 beginning with T.

Since $V_x = 2 Q x + q$, the optimal control function u^* is now obtained from (13.2.2):

$$u^*(s, x) = -\frac{1}{2} D(s)^{-1} B(s)' (2 Q(s) x + q(s)).$$

For $a(T) = 0$, we have $q(s) \equiv 0$ and hence

(13.2.5) $$u^*(s, x) = -D(s)^{-1} B(s)' Q(s) x.$$

13.3 Control on the Basis of Filtered Observations

A control function $u(t, X_t) = U_t$ depends on the state X_t of the system, which we have up to now assumed to be known exactly. However, in many cases, only noisy observations of a function of X_t are possible, so that the control and filtering must be combined. In the linear case, this yields

$$dX_t = A(t) X_t \, dt + B(t) U_t \, dt + G(t) \, dW_t,$$

(13.3.1)

$$X_{t_0} = c \ \Re(0, E c c')\text{-distributed},$$

(13.3.2) $\quad dZ_t = H(t) X_t \, dt + R(t) \, dV_t$

where X_t, U_t, Z_t, W_t and V_t assume values in Euclidean spaces of arbitrary dimensions, W_t, V_t and c are independent, and $R(t) R(t)'$ is positive-definite. If we fix the cost functional (13.1.2) by

$$k(t, x, u) = x' C(t) x + u' D(t) u,$$

$$M(T, x) = x' F(T) x,$$

where $C(t)$ and $F(t)$ are symmetric and nonnegative-definite and $D(t)$ is symmetric and positive-definite, then the control and filtering can be separated from each other. In the following discussion, we shall follow Bucy and Joseph ([61], pp. 96–102).

Since instead of X_t we know only the observations Z_t, our control functions u are allowed to depend not on X_t but only on the observations $Z[t_0, t]$; that is, we are considering functionals of the form

$$U_t = u(t, Z[t_0, t]).$$

Also, the cost functional must now be considered under condition of a certain observation:

$$V^u(s, \widehat{X}_s) = E\left((X_T^u)' F(T) X_T^u + \int_s^T ((X_r^u)' C(r) X_r^u + U_r' D(r) U_r) \, dr \, | \, Z[t_0, s]\right).$$

Here, $\widehat{X}_s = E(X_s^u | Z[t_0, s])$ and X_t^u is a solution of (13.3.1). Then, there exists an optimal control function, namely,

(13.3.3) $\quad U_t^* = -D(t)^{-1} B(t)' Q(t) \widehat{X}_t.$

Here, $Q(t)$ is the symmetric solution of

(13.3.4) $\quad \dot{Q}(t) + A' Q + Q A - Q B D^{-1} B' Q + C(t) = 0$

on the interval $[t_0, T]$ with the end condition $Q(T) = F(T)$ (see equation (13.2.4)), and $\widehat{X}_t = E(X_t | Z[t_0, t])$ is the solution of

(13.3.5) $\quad d\widehat{X}_t = A X_t \, dt + B U_t^* \, dt + P H' (R R')^{-1} (dZ_t - H \widehat{X}_t \, dt)$

on the interval $[t_0, T]$ with the initial value $\widehat{X}_{t_0} = 0$. Finally, $P(t) = E\,((X_t - \widehat{X}_t)\,(X_t - \widehat{X}_t)'\,|\,Z\,[t_0, t])$ is the error covariance matrix which is independent of the control function and of the observation and which satisfies the equation

$$\dot{P}(t) = A\,P + P\,A' - P\,H'\,(R\,R')\,H\,P + G\,G', \quad t_0 \leqq t \leqq T,$$

$$P(t_0) = E\,c\,c',$$

The minimum costs arising from use of the control function U_t^* with the estimated starting point $\widehat{X}_s$ in $[s, T]$ are then

$$V(s, \widehat{X}_s) = \widehat{X}_s'\,Q(s)\,\widehat{X}_s + \int_s^T \mathrm{tr}\,(P\,H'\,(R\,R')^{-1}\,H\,P\,Q)\,\mathrm{d}r$$

$$+\,\mathrm{tr}\,F(T)\,P(T) + \int_s^T \mathrm{tr}\,(C(r)\,P(r))\,\mathrm{d}r.$$

These equations are said to contain the so-called **separation principle**:

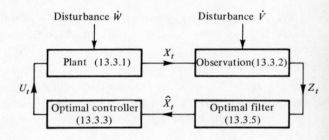

Fig. 10:
Separation of filtering and control.

The combined filtering and control problem can obviously be broken into the following problems:

1. Filtering: determination of the optimal estimate $\widehat{X}_t$ of X_t on the basis of the observation $Z\,[t_0, t]$ from equation (13.3.5).

2. Determination of the optimal control function $u^* = u^*\,(t, X_t)$ for the *deterministic* problem $(G \equiv 0)$. We get

$$u^* = -D(t)^{-1}\,B(t)'\,Q(t)\,X_t,$$

where Q is the solution of equation (13.3.4) (cf. (13.2.5)). The optimal control function for the stochastic problem is then simply

$$U_t^* = u^*\,(t, \widehat{X}_t),$$

and hence is in fact a functional of the observations $Z\,[t_0, t]$.

Bibliography

1. **Selection of Texts on Probability Theory and the Theory of Stochastic Processes**

[1] Bauer, H., *Wahrscheinlichkeitstheorie und Grundzüge der Masstheorie I*, Berlin, de Gruyter 1964 (Sammlung Göschen, Bd. 1216/1216a).

[2] Bauer, H., *Wahrscheinlichkeitstheorie und Grundzüge der Masstheorie*, Berlin, de Gruyter 1968.

[3] Doob, J., *Stochastic Processes*, New York: Wiley 1960.

[4] Feller, W., *An Introduction to Probability Theory and its Applications*, Vols. 1, 2, New York, Wiley 1955/1966.

[5] Gikhman, I. I., and Skorokhod, A. V., *Introduction to the Theory of Random Processes*, Philadelphia, W. B. Saunders, 1969 (translation from Russian).

[6] Hinderer, K., *Grundbegriffe der Masstheorie und der Wahrscheinlichkeitstheorie*, Ausarbeitung einer Vorlesung an der Universität Hamburg, 1968/1969.

[7] Krickeberg, K., *Wahrscheinlichkeitstheorie*, Stuttgart, Teubner, 1963.

[8] Lamperti, J., *Probability*, New York, Benjamin, 1966.

[9] Loève, M., *Probability Theory*, Princeton, Van Nostrand, 1953.

[10] Meyer, P. A., *Probability and Potentials*, Waltham, Mass., Blaisdell, 1966.

[11] Morgenstern, D., *Einführung in die Wahrscheinlichkeitsrechnung und mathematische Statistik*, Berlin, Göttingen, and Heidelberg, Springer, 1964.

[12] Neveu, J., *Mathematical Foundations of the Calculus of Probability*, San Francisco, Holden-Day, 1965.

[13] Papoulis, A., *Probability, Random Variables, and Stochastic Processes*, New York, McGraw-Hill 1965.

[14] Prabhu, N. V., *Stochastic Processes*, New York, MacMillan, 1965.

[15] Prohorov, Yu. V., and Rozanov, Yu. A.: *Probability Theory*, Berlin, Heidelburg, and New York, Springer, 1969 (translation from Russian).

[16] Renyi, A., *Foundations of Probability*, San Francisco, Holden-Day, 1970.

[17] Richter, H., *Wahrscheinlichkeitstheorie*, Berlin, Göttingen, and Heidelberg, Springer, 1956.

2.   Markov and Diffusion Processes, Wiener Processes, White Noise

See the corresponding sections in [2], [3], [4], [5], [7], [8], [9], [10], [12], [13], [14], [15], [45].

[18]   Bauer, H., *Markoffsche Prozesse,* Ausarbeitung einer Vorlesung an der Universität Hamburg, 1963.

[19]   Bharucha-Reid, A. T., *Elements of the Theory of Markov Processes and their Applications,* New York, McGraw-Hill, 1960.

[20]   Dynkin, E. B., *Theory of Markov Processes,* Englewood Cliffs, New Jersey, Prentice-Hall, 1961 (translation from Russian).

[21]   Dynkin, E. B., *Markov Processes,* Vol. 1, 2. Berlin, Göttingen, and Heidelberg, Springer, 1965 (translation from Russian).

[22]   Gelfand, I. M.; Vilenkin, N. J., Generalized Functions, vol. 4, New York, Academic Press, 1961 (translation from Russian).

[23]   Hunt, G. A., *Martingales et processus de Markov,* Paris, Dunod, 1966.

[24]   Itô, K.: *Lectures on Stochastic Processes,* Bombay, Tata Institute of Fundamental Research, 1961.

[25]   Itô, K.: *Stochastic Processes,* Aarhus, Universitet, Matematisk Institut, 1969 (Lecture Notes Series, No 16).

[26]   Itô K., and McKean, H. P., *Diffusion Processes and their Sample Paths,* Berlin, Heidelberg, and New York, Springer, 1965.

[27]   Levy, P., *Processus stochastiques et mouvement brownien,* Gauthier-Villars, 1948.

[28]   Mandl, P., *Analytical Treatment of One-Dimensional Markov Processes,* Berlin, Heidelberg, and New York, Springer, 1968.

[29]   Nelson, E., *Dynamical Theories of Brownian Motion,* Princeton University, Press, 1967.

3.   Stochastic Differential Equations

See the corresponding sections in [3], [5], [21], [25], [29], [61], [65], [66], [72], [80].

[30]   Anderson, W. J., *Local Behaviour of Solutions of Stochastic Integral Equations,* Ph.D. thesis, Montreal, McGill University, 1969.

[30a] Arnold, L., "The loglog law for multidimensional stochastic integrals and diffusion processes," *Bull. Austral. Math. Soc.,* 5 (1971), pp. 351-356.

[31]   Bharucha-Reid, A. T., *Random Integral Equations,* New York, Academic Press (in press).

[32]   Chandrasekhar, S., "Stochastic problems in physics and astronomy," *Rev. Mod. Phys.,* 15 (1943), pp. 1-89 (contained in [51]).

[33] Clark, J. M. C., *The Representation of Nonlinear Stochastic Systems with Application to Filtering*, Ph. D. thesis, London, Imperial College, 1966.

[34] Dawson, D. A., "Generalized stochastic integrals and equations," *Trans. Amer. Math. Soc.*, 147 (1970), pp. 473-506.

[35] Doob, J. L., "The Brownian movement and stochastic equations," *Ann. Math.*, 43 (1942), pp. 351-369 (contained in [51]).

[36] Gikhman, I. I., and Skorokhod, A. V.: Stochastische Differentialgleichungen, Berlin, Akademie-Verlag, 1971 (translation from Russian).

[37] Girsanov, I. V., Primer neyedinstvennosti resheniya stokhasticheskogo uravneniya K. Ito (an example of nonuniqueness of the solution of K. Itô's stochastic equation), Teoriya veroyatnostey i yeye primeneniya, 7 (1962), pp. 336-342.

[37a] Goldstein, J. A., "Second order Itô processes," *Nagoya Math. J.*, 36 (1969), pp. 27-63.

[38] Gray, A. H., *Stability and Related Problems in Randomly Excited Systems*, doctoral thesis, Pasedena, California Institute of Technology press, 1964.

[39] Gray, A. H., and Caughey, T. K., "A controversy in problems involving random parametric excitation," *J. Math. and Phys.*, 44 (1965), pp. 288-296.

[40] Itô, K., "Stochastic differential equations in a differentiable manifold," *Nagoya Math. J.*, 1 (1950), pp. 35-47.

[41] Itô, K., "On a formula concerning stochastic differentials," *Nagoya Math. J.*, 3 (1951), pp. 55-65.

[42] Itô, K., *On Stochastic Differential Equations*, New York, Amer. Math. Soc., 1951 (Memoirs, Amer. Math. Soc. No. 4).

[43] Itô, K., and Nisio, M., "On stationary solutions of stochastic differential equations," *Journ. of Math. of Kyoto Univ.*, 4 (1964), pp. 1-79.

[44] Langevin, P.: "Sur la théorie du mouvement brownien," *C. R. Acad. Sci.*, Paris, 146 (1908), pp. 530-533.

[45] McKean, H. P.: *Stochastic Integrals*, New York, Academic Press, 1969.

[46] McShane, E. J., "Toward a Stochastic Calculus," *Proc. Nat. Acad. Sci.*, USA, 63 (1969), p. 275-280, and 63 (1969), pp. 1084-1087.

[47] Skorokhod, A. V., *Studies in the Theory of Random Processes*, Reading, Mass., Addison-Wesley, 1965 (translation from Russian).

[48] Stratonovich, R. L., "A new representation for stochastic integrals and equations," *SIAM J. Control*, 4 (1966), pp. 362-371.

[49] Uhlenbeck, G. E., and Ornstein, L. S., "On the theory of Brownian motion," *Phys. Rev.*, 36 (1930), pp. 823-841 (contained in [51]).

[50] Wang, M. C., and Uhlenbeck, G. E., "On the theory of Brownian motion II," *Rev. Mod. Phys.*, 17 (1945) pp. 323-342 (contained in [51]).

[51] Wax, N. (ed.), *Selected Papers on Noise and Stochastic Processes,* New
       York, Dover, 1954, (contains [32], [35], [49], [50]).

[52] Wong, E., and Zakai, M., "On the convergence of ordinary integrals to sto-
       chastic integrals," *Ann. Math. Statist.,* 36 (1965) pp. 1560-1564.

[53] Wong, E., and Zakai, M., "The oscillation of stochastic integrals," *Z. Wahrs-
       cheinlichkeitstheorie verw. Geb.,* 4 (1965) pp. 103-112.

[54] Wong, E., and Zakai, M.: "Riemann-Stieltjes approximation of stochastic
       integrals," *Z. Wahrscheinlichkeitstheorie verw. Geb.,* 12 (1969), pp.
       87-97.

4.   Stability, Filtering, Control

[55] Aoki, M., *Optimization of Stochastic Systems,* New York, Academic Press,
       1967.

[56] Athans, M., and Falb, P. L., *Optimal Control, An Introduction to the
       Theory and its Applications,* New York, McGraw-Hill, 1966.

[57] Bellman, R., *Dynamic Programming,* Princeton University Press, 1957.

[58] Bhatia, N. P., and Szegö, G. P., *Stability Theory of Dynamical Systems,*
       Berlin, Heidelberg, and New York, Springer, 1970.

[59] Bucy, R. S., "Stability and positive supermartingales," *J. Differential Equ.,*
       1 (1965), pp. 151-155.

[60] Bucy, R. S., "Nonlinear filtering theory," *IEEE Trans. Automatic Control,*
       10 (1965), p. 198.

[61] Bucy, R. S., and Joseph, P. D., *Filtering for Stochastic Processes with Ap-
       plications to Guidance,* New York, Interscience Publ., 1968.

[62] Fleming, W. H.: "Optimal continuous-parameter stochastic control," *SIAM
       Review,* 11 (1969), pp. 470-509.

[63] Gikhman, I. I., "On the stability of the solutions of stochastic differential
       equations" (in Russian), *Predel'nyye teoremy i statisticheskiye vyvody,*
       Tashkent, 14-45, 1966.

[64] Hahn, W., *Stability of Motion,* Berlin, Heidelberg, and New York: Springer,
       1967.

[65] Khas'minskiy, R. Z., Ustoychivost' sistem differentsial'nykh uravneniy pri
       sluchaynykh vozmushcheniyakh (Stability of systems of differential
       equations in the presence of random disturbances), Moscow, Nauka
       press, 1969.

[66] Jazwinski, A. H., *Stochastic Processes and Filtering Theory,* New York,
       Academic Press, 1970.

[67] Kalman, R. E.: New methods in Wiener filtering theory, Proc. First Sympos.
       on Engin. Appl. on Random Function Theory and Probability (edit.
       by J. L. Bogdanoff and F. Kozin), New York, Wiley 1963, pp. 270-388.

[68] Kalman, R. E., Falb, P. L., and Arbib, M. A., *Topics in Mathematical System Theory*, New York, McGraw-Hill, 1969.

[69] Kolmogorov, A. N., Interpolirovaniye i ekstrapolirovaniye statsionarnykh sluchaynykh posledovatel'nostey (Interpolation and extrapolation of stationary random sequences), Izvestiya Akad. nauk (seriya matematicheskaya), 5 (1941), pp. 3-14.

[70] Kozin, F., "On almost sure asymptotic sample properties of diffusion processes defined by stochastic differential equations," *Journ. of Math. of Kyoto Univ.*, 4 (1965), pp. 515-528.

[71] Kushner, H. J., "On the differential equations satisfied by conditional probability densities of Markov processes," *SIAM J. Control*, 2 (1964), pp. 106-119.

[72] Kushner, H. J., *Stochastic Stability and Control*, New York, Academic Press, 1967.

[73] Morozan, T., Stabilitatea sistemelor cu parametri aleatori (stability of systems with random parameters), Bucharest, Editura Academiei Republicii Socialiste România, 1969.

[74] Mortensen, R. E., *Optimal Control of Continuous-Time Stochastic Systems*, Ph.D. thesis (engineering), Berkeley, Univ. of California press, 1966.

[75] Sagirow, P., *Stochastic Methods in the Dynamics of Satellites*, Lecture Notes, Udine, CISM, 1970.

[76] Stratonovich, R. L.: *Topics in the Theory of Random Noise*, Vol. 1, New York, Gordon and Breach, 1963 (translation from Russian).

[77] Stratonovich, R. L.: *Conditional Markov Processes and Their Application to the Theory of Optimal Control*, New York, American Elsevier, 1968 (translation from Russian).

[78] Strauss, Aaron: *An Introduction to Optimal Control Theory*, Berlin, Heidelberg, and New York, Springer, 1968 (Lecture notes in operations research and mathematical economics, Vol. 3.)

[79] Wiener, N.: *Extrapolation, Interpolation and Smoothing of Stationary Time Series with Engineering Applications*, Cambridge, Mass., Mass. Inst. Tech. press, 1949.

[80] Wonham, W. M., "Random differential equations in control theory" in A. T. Bharucha-Reid (ed.): *Probabilistic Methods in Applied Mathematics*, Vol. 2, New York, Academic Press, 1970, pp. 131-212.

Subject Index

DATE DUE